ASTRONOMY
LIBRARY

THE MOMENT OF CREATION

Also by James S. Trefil

FROM ATOMS TO QUARKS
ARE WE ALONE? (with Robert T. Rood)
LIVING IN SPACE
THE UNEXPECTED VISTA

THE MOMENT OF CREATION

Big Bang Physics From Before the First Millisecond to the Present Universe

JAMES S. TREFIL

Illustrations by Gloria Walters

CHARLES SCRIBNER'S SONS • NEW YORK

This book is dedicated to the following men, each of whom labored to teach me a part of what he knew:

Wallar Meuhl, J. Sterling Morton High School
Robert Hulsizer, University of Illinois
Rom Harré, Oxford University
Sidney Drell, Stanford University

Library of Congress Cataloging in Publication Data

Trefil, James S., 1938–
The moment of creation.

Includes index.
1. Big bang theory. 2. Cosmology—History.
3. Nuclear astrophysics. I. Title.
QB991.B54T7 1983 523.1'8 83-9011
ISBN 0-684-17963-6

This book published simultaneously
in the United States of America and in Canada—

3 5 7 9 11 13 15 17 19 F/C 20 18 16 14 12 10 8 6 4

Printed in the United States of America.

Contents

Introduction

"Where did it all come from?" There are few questions that grip the mind more forcefully than those about the creation of the universe. Every society has given a prominent place in its folklore to an explanation of how the world came to be. Originally these creation epics revolved around the deeds of gods whose whims, battles, or needs caused the world and its human inhabitants to be created. But even as long ago as the sixth century B.C., philosophers in the Greek colonies along the eastern Mediterranean were trying to explain how the elements of earth, fire, air, and water could interact to produce the cosmos as we know it. With those philosophers, the emphasis on the search for understanding shifted from "Who?" to "How?"

Western civilization's contribution to this long-standing human endeavor is impressive and, as we shall shortly see, has become much more impressive in the last few years. Since the early part of the twentieth century, we have known that there is a general expansion in the universe, with distant galaxies receding from our own Milky Way. As our ability to measure the properties of distant parts of the universe has grown, a picture has emerged—a picture that constitutes the creation epic as told by the scientific method. The weight of the accumulated evidence tells us that the expansion we see is the result of a titanic explosion that took place about 15 billion years ago, an explosion we call the Big Bang.

One way of measuring the completeness of our knowledge of the history of the universe is to ask how far back we can trace the evolution of matter, or, to put it another way, how close we can get to the actual moment of creation in the Big Bang. In the early years of this century, we knew about the law of universal gravitation, so we could discuss the formation of the galaxies (and stars within the galaxies) as an example of gravity acting to pull large masses together. This meant we could get to within about 500,000 years of the Big Bang. By the 1940s, we had learned enough about the atomic nucleus and the more common elementary particles to take another giant step. We could now describe the process by which protons and neutrons came together to form the nuclei of hydrogen and helium, the raw material of the stars. This brought us to within about three minutes of the moment of creation.

By the mid-1960s, our increasing insight into the structure and behavior of elementary particles—an insight gained through a tremendous effort on the part of a large number of experimental and theoretical physicists—allowed us to trace out the somewhat convoluted evolution of the particles that eventually condensed into nuclei. By 1970 we had pushed the boundaries to knowledge to within about 1 millisecond (10^{-3} second) of the Big Bang.

It would seem that this level of knowledge would be enough to satisfy anyone, but even as cosmologists were sorting out the details of the postmillisecond period, theoretical physicists were laying the groundwork for the next major advance. Spurred by unresolved problems in the field of particle physics, a new type of theory was born, one that resulted in an immense simplification of our picture of the seemingly complex subatomic world. Called *gauge theories,* these new ideas swept through the community of physicists, bringing with them an entirely new way of looking at natural phenomena. Because these theories are so far-reaching, it was almost inevitable that they would be used to push the boundary between the known and the unknown still farther back toward the moment of creation.

During the past few years, the most exciting area in science has been in this overlapping area between high-energy physics and astronomy. The frontier of knowledge of the early universe has been pushed back to somewhere between 10^{-35} and 10^{-43} second of the Big Bang—an interval so short that our concept of time itself may have to be redefined before we go any further. This synthesis has been so successful that theorists are now engaged in an attempt to push the frontier all the way back to the moment of creation itself, and a few of the more audacious investigators are going further and asking the ultimate question: Why does the universe exist at all?

The Moment of Creation tells the story of this enormous explosion of our knowledge of the early universe. After a short historical introduction to bring us back to the first millisecond, the new physics of gauge theories is explained. Because some of the concepts in these theories may be difficult, there is a fast track provided for the reader who wants to get on with cosmology. At the end of each chapter in Part II is a summary that supplies the essential facts about the theories, without going into the justification for those facts. Once the new physics is at our disposal, we will resume our march toward the moment of creation until we come to the current frontier, at which point we will try to guess what may lie beyond it.

There is very little mathematics in this book, but I frequently have to use very large and very small numbers to describe important quantities. For these I have used the conventional shorthand of the so-called scientific notation. In this system, all numbers are written as a number between 1 and 10 multiplied by a power of 10. If the power of 10 is positive, we interpret it as an instruction to move the decimal point a certain number of places to the right. Thus, 2×10^2 would be 200, and 3×10^6 would be 3,000,000 (3 million). If the power of 10 is negative, we interpret it as an instruction to move the decimal point to the left. Thus, 2×10^{-3} is .002, and 10^{-35} is a decimal point followed by thirty-four zeros and a 1. This notation will save us from carrying around a lot of excess baggage in the form of zeros.

Finally, I would like to thank a number of people who helped me with this book. Gloria Walters produced the illustrations on a very tight schedule and Nancy Lane was, as usual, invaluable in the process of putting the manuscript together. The number of my colleagues whose comments and conversations added to the book is too large to permit me to name each one, but I would particularly like to thank Paul Fishbane and Robert Rood of the University of Virginia for their assistance, with the usual statement that I alone am responsible for any errors that might remain in the final product.

JAMES TREFIL
Charlottesville, Virginia
May 1983

PART ONE

Since the discovery of the expansion of the universe in the 1920s, one goal of cosmology has been to trace the history of the universe—in effect, to run the film backward until we can see and understand how the universe came into being. A measure of our success in this endeavor is the brevity of the time that separates the frontier of our knowledge from the moment of creation. In the 1920s, this time was about 3 minutes. Until a few years ago, before the advances that are the main topic of this book, it was 1 millisecond. Now it may be as small as 10^{-35} second.

In this section of the book we trace the history of discovery concerning the Big Bang back to within the first millisecond and discuss some of the problems that the conventional Big Bang ideas present. We shall see that if we expect to solve these problems, we are forced to think about what happened in the universe during the first millisecond of its existence.

Chapter 1

The Big Bang

One thing about the past.
It's likely to last.
OGDEN NASH

It took us several thousand years to discover that the solid earth on which we live is not the center of the universe, the unmoved core around which the rest of the heavens revolve. But after Copernicus and Kepler taught us that the earth moved and the sun stood still, it did not take us long to realize that the stars were bodies much like the sun. The fact that they appeared fixed in the heavens was taken (rightly) to indicate that they lie very far from our solar system. By 1750, only a little more than a century after Galileo had been forced to recant the Copernican doctrine at his trial for heresy in 1633, the English astronomer Thomas Wright was putting forward the idea that the appearance of the Milky Way was evidence that the stars near the solar system are arranged in a flat, disklike structure. We now call this structure the Milky Way galaxy (*galacticos* actually means "milky" in Greek). It is a collection of some 10 billion stars, of which our own sun is a rather ordinary member. How many other planetary systems there may be remains unknown.

If the sun is part of a collection of stars arranged in a certain way, it is natural to ask whether other such collections exist in the universe. Debate on this question went on throughout the late eighteenth and nineteenth centuries. It centered on faint patches of light in the sky known as nebulae, of which there are many. One, in the constellation Andromeda, had been known to Arab and Persian astronomers since the tenth century. The famous astronomer Sir William Herschel suggested that some of these nebulae might be galaxies like the Milky Way, as did Immanuel Kant. Yet there are many nebulae in the sky that are not galaxies. There are, for example, clouds of gas illuminated by stars in their interior, presenting the same appearance as galaxies. Such nebulae are clearly in our own galaxy, and strong arguments were made against assigning huge distances to any other nebulae in the sky. One argument centered on the fact that flashes of light were occasionally seen originating in some nebulae—flashes of such brightness that no energy source known to nineteenth-century astronomers could have produced them unless the distance to the origin of the flash was relatively small. Today we know that these flashes were supernovae and that the source of energy is the conversion of huge amounts of matter into energy. In the nineteenth century, however, supernovae were not understood and the conversion of mass to energy was unknown, so the question of whether some nebulae were indeed distant galaxies or merely clouds within the Milky Way remained very much open.

It was not until 1923, after the 100-inch telescope at the Mount Wilson Observatory in the mountains above Los Angeles was completed, that Edwin Hubble was able to resolve this puzzle. Because of the power of the new instrument, it was possible to pick out individual stars in the Andromeda nebula. The types of stars Hubble saw were similar to those in our own galaxy; in particular, he was able to isolate several stars of the type known as Cepheid variables. These are stars whose brightness varies periodically on a time scale of weeks or months, and it was known that the period of the variation was related to the total amount of light emitted by the star. This fact can be used to measure distances in our own galaxy, because if we know how much light a star is emitting and how much we actually receive on the earth, some simple arithmetic will tell us how far away the star must be. Using this well-known technique on the stars observed in the Andromeda nebula, Hubble was able to show that the stars (and hence the nebula) must be almost a million light-years away. Since the Milky Way is only 100,000 light-years across, the only possible conclusion from Hubble's work was that the Milky Way must be only one of many galaxies in the universe. Since Hubble's time, the number of known galaxies has passed the million mark, and the growth of this number shows no sign of slowing down.

Just as those who followed Copernicus had to surrender the

concept of the earth and the solar system as the center of the universe, scientists in the twentieth century have had to abandon the concept of the uniqueness of the Milky Way. It turns out that we live in a rather typical galaxy, with nothing to distinguish it from others except for the fact that it is our home.

The Universal Expansion

Once the existence of other galaxies was firmly established, Hubble's continued studies revealed another important (and, at the time, surprising) thing. It appeared that the distance between the Milky Way and other galaxies was increasing—that other galaxies were moving away from us. Furthermore, it appeared that the farther away from us the galaxy in question was, the faster was the speed of its recession. So crucial is this observation to the development of modern cosmology that it is worth spending some time understanding how Hubble arrived at this result.

Anyone who has stood next to a railroad track and listened to the whistle of a passing train knows that the pitch changes as the train approaches and then passes into the distance. As the train approaches us, the pitch of the whistle grows higher, and as the train moves away, the pitch drops. This is a specific example of a rather general phenomenon associated with waves of all kinds, a phenomenon known as the Doppler effect. Figures 1 and 2 can help us to understand the effect.

The source of the sound wave (the train whistle in our example) creates a repeating pattern of pressure in the air. When these pressure patterns reach the ear, they vibrate the eardrum and we hear a sound. The more high-pressure ridges that strike the ear each second, the faster the eardrum vibrates and the higher the pitch of the sound we perceive. If the source is stationary, as in Figure 1, then the waves move out in uniformly spaced concentric spheres. Any observer will hear the same pitch, no matter where he stands.

If, on the other hand, the source of sound is moving (as with our train), then the situation is different. Each wave will go out as a sphere, but that sphere will be centered on the position of the source at the time the wave was emitted. This means that the spheres will no longer be concentric but will bunch up, as shown in Figure 2. An observer standing in front of the source at *A* will find the crests of the wave more closely spaced than normal, and

Hubble

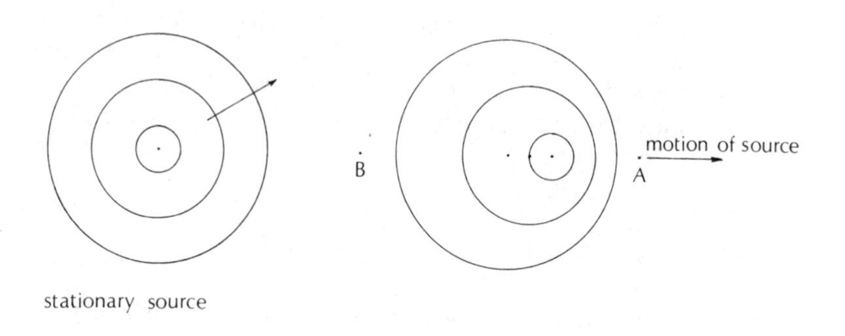

Figure 1. *Figure 2.*

he will hear a higher-pitched sound. Conversely, someone standing in back of the source at *B* will find the crests more widely separated and will hear a lower sound. This explains the train whistle phenomenon.

This effect was first suggested by mathematician Christian Johann Doppler at Prague in 1842 and was verified experimentally three years later by the Dutch scientist Christoph Buys Ballot. His experiment was as direct as it was picturesque. He assembled a group of musicians who had perfect pitch, and had a group of trumpeters pass by on an open railway car. Compared to this technique, our modern testing methods that employ sensitive microphones and oscilloscopes seem hopelessly prosaic.

Light and sound are both wave phenomena, and it is clear that Doppler's argument will apply to both, even though the nature of the waves is quite different in the two cases. What this means is that if we look at light emitted by a particular atom in a distant galaxy and then look at light emitted by that same species of atom in our laboratory, a comparison of the perceived wavelengths will tell us whether that galaxy is moving toward us or away from us. In the case of light, the bunching up of the wave crests seen at point *A* in Figure 2 will be perceived as a shift toward the blue, or high-frequency, end of the spectrum, while the spreading out at point *B* will be perceived as a shift toward the red, or low-frequency, end.

What Hubble found was that every galaxy whose distance he could determine with his instruments showed a red shift, indicating that every galaxy is moving away from us. Furthermore, Hubble concluded from his data that the farther away the galaxy was (with distance measured by the Cepheid variable data), the higher was its red shift.

The picture that emerged from Hubble's work, therefore, was of a universal expansion of the universe, with all galaxies receding from each other. Knowing the separation of the galaxies at the present time and the speed at which they are receding, it is a simple exercise to imagine "running the film in reverse" and letting the galaxies get closer and closer together. In principle, then, we could infer from the present condition of the galaxies the amount of time that has elapsed since all the matter in the universe existed in a highly compact state. The number we get in this way is called the Hubble time or the Hubble age and represents a rough guess at the age of the universe. This particular number has a long and somewhat checkered history.

Measuring the Hubble age requires that we have a very good understanding of the Cepheid variable stars and that we have very accurate measurements of the red shift of other galaxies. By the nature of things, those galaxies that are farthest away will have the largest red shift, and it is for these galaxies that we will expect the most accurate measurements of the Doppler effect. On the other hand, it is precisely for these galaxies that the Cepheid variables will be hardest to distinguish. The inverse situation holds for nearby galaxies. In other words, it is not a trivial matter to measure both the red shift and the distance with sufficient accuracy to make a precise determination of the Hubble age. When Hubble made the first measurement in 1931, he set the age of the universe at only a billion years or so—about ten times smaller than the presently accepted number. The problem with Hubble's billion years was that the age of the earth as determined by geologists was thought to be several billion years, and the earth could hardly be older than the universe. Improved techniques and a recalibration of the Cepheid-variable scale have led to repeated changes in the best estimate of the Hubble age, and even today astronomers can make headlines by announcing that the universe is only half as old (or twice as old) as had been previously thought. Creationists, as might be expected, enjoy pouncing on such statements as evidence against the entire evolutionary picture of the universe. A much more reasonable attitude is that of one astronomer whose judgment I have learned to trust: "The Hubble age is 15 billion years, plus or minus 50 percent." In other words, the best guess for the Hubble age is 15 billion years, a number we will adopt for the rest of this book. We cannot, however, rule out ages as low as 7 billion or as high as 20 billion years. Since this uncertainty is not terribly important to most modern cosmology, a better determination of the Hubble age is not regarded as an important goal in astronomy.

The only force that now acts on an outward-moving galaxy is the gravitational attraction of the rest of the galaxies. We know enough about the way matter behaves under the influence of gravity to be able to trace the expansion backward in time. We cannot believe, however, that we can use this sort of simple extrapolation all the way back to the Big Bang. At the early stages of the expansion, the density of matter would have been high enough so that electrical and even nuclear forces could be expected to play a role in the development of the universe. This is not a trivial statement, since our theories tell us that the present uniform expansion need not be typical of earlier epochs.

The Big Bang

Despite these uncertainties in the exact age of the universe, it has become accepted that the expansion that we see now began about 15 billion years ago, from a situation in which all matter was in a highly condensed state. The term *Big Bang* is applied to this event, and the picture this term conveys of a universe consisting of fragments hurled out from an exploding center is essentially accurate. The only problem with this simple view of things is that it seems to imply that the Big Bang was analogous to the explosion of an artillery shell, with fragments flying out through space. Actually, modern cosmologists have a much more complex view of the process.

The theory of relativity teaches us that there is a delicate interconnection between matter and space. If we think of free space as a kind of grid marked out in the void, then matter has the effect of distorting the grid. In this view, the expansion of matter that we call the Big Bang would carry the grid with it, stretching it out as if it were a rubber sheet. A way of visualizing the expansion as it is seen by a modern cosmologist is to think of blowing up a balloon that has dots scattered over its surface, with each dot representing a galaxy. To someone on one of the dots, it is not so much a question of the balloon expanding through space

as it is an expansion of space itself (as represented by the balloon's surface).

There is another important point about the universal expansion that can be understood in terms of the balloon. If you stood on one dot and looked around you, it would appear to you as if you were standing still and every other dot were receding from you. In other words, it would appear to you as if you were the center of the universal expansion, occupying a privileged place in the universe. This would not be true, however, because it is clear that an observer on any other dot would see exactly the same thing. Everyone sees himself as being stationary and everyone else receding, so no observer's view of the universe is any more privileged than that of anyone else.

This is a special case of the principle of relativity. In its most general form, this principle states that the laws of nature must be such that every observer will find them to be identical—that there is no privileged frame from which to observe the universe. The principle, so compatible with the working of the twentieth-century mind, is built into contemporary physics from the foundation up. In a sense, it plays a role in modern science similar to the geocentrism of ancient and medieval sciences. We build it into our perceptions of the universe because we feel it ought to be true, and, as yet, we have found no experimental evidence to convince us that it is not. But then, geocentrism worked beautifully for several millennia, too.

Echoes

Although the idea that the universe may have originated in a cataclysmic explosion billions of years ago is implicit in Hubble's discoveries, these discoveries do not force us to accept the Big Bang.

Cosmologists are marvelously imaginative people, able to produce alternative explanations for any fact at the drop of a hat. For a long time, the chief rival to the Big Bang picture was a theory in which, as the galaxies receded from each other, new matter was created in the voids, creating new galaxies that would, in turn, recede and be replaced by still newer ones. The simple existence of the red shift data in and of themselves could not distinguish between the Big Bang and this alternative. Before the Big Bang could be accepted, some more direct confirmation would have to be found.

We can understand what this direct evidence turned out to be if we concentrate on the fact that in its earlier stages of development the universe must have been very hot. Physicists know that any object whose temperature exceeds absolute zero radiates energy and that the type of energy radiated depends on the object's temperature. Anyone who has watched the heating element on an electric stove turn from dull red to bright orange has firsthand experience to corroborate this statement. The element is giving off that particular type of radiation we call visible light, radiation to which our eyes are sensitive and which can therefore be seen directly.

Even after the heating element has been turned off and its color has returned to normal, it is still radiating. You can verify this by putting your hand near the stove. You will feel heat coming out, even though there is no visible radiation. What has happened is that as the temperature of the heating element has dropped, the radiation given off has moved out of the visible range and into the infrared. Our eyes are not sensitive to this radiation (although it can readily be detected by instruments and special photographic film). Nevertheless, we can perceive the radiation with our hand. The colder the object being observed, the farther away from being visible is the radiation, and the more elaborate will our detector have to be for us to sense it.

The universe was very hot a long time ago, and it has been expanding and cooling off ever since. Like the heating element on the stove, the radiation given off during this cooling process has shifted from very energetic to less energetic and has long since passed out of the range associated with visible light. In fact, as long ago as 1948, theoretical physicists, taking into account the rate of expansion of the universe, had predicted that 15 billion years after the Big Bang the universe should have cooled off to a temperature just 3° above absolute zero, a number usually referred to as 3° on the Kelvin scale, or 3°K. Even at such an extremely low temperature, the laws of physics tell us that a body gives off radiation. Furthermore, they tell us that the radiation will be in the microwave range. Microwave radiation is intermediate between radio waves and infrared radiation in energy; it is the same radiation that cooks food in a microwave oven and is emitted by aircraft radar systems. This radiation should be bathing the entire volume of the universe by now, and we should see it coming to the earth, no matter in which direction we look. In this respect, we are in a position analogous to an observer surrounded by a fire: he feels heat coming toward him from every direction.

There is another way to think of this prediction. During the early stages of the Big Bang, when the universe was very hot, radiation could not travel very far without encountering some sort of matter. These continual collisions had the effect of ensuring that the radiation was of the type characteristic of the general temperature of the universe at that time. In practical terms, it guaranteed that the radiation would be primarily X rays or gamma rays. Then, about 500,000 years after the Big Bang, the universe suddenly underwent a change that had the effect of lessening the probability that radiation would collide with matter. (This very important transition will be discussed in detail in the next chapter.) We can think of the radiation after this time as being analogous to a gas that is confined, heated to a high temperature, and then suddenly released. It will expand outward with the rest of the universe, cooling off as it does so. It is not difficult to see that it is possible to predict the temperature of the radiation gas at any time in the future, and this is precisely where the 3°K comes from. It is the temperature to which the radiation gas would have dropped by the present time.

Thus, whether we think of the radiation as being characteristic of an expanding universe or as an expanding gas that stopped interacting with matter long ago at a high temperature is irrelevant. In either case, we come to the conclusion that if a universal microwave radiation did exist in the universe, it would be a direct way of perceiving the early stages of the Big Bang, in much the same way that light emitted from a distant star allows us to see that star and determine its condition. It would be pleasant to report that a neat sequence of prediction, experiment, and confirmation led to the acceptance of this idea, but as so often happens in science, the actual chain of events was not nearly so simple.

In the early 1960s, the first primitive communication satellites were being launched. The Echo series were little more than reflecting spheres in orbit, designed to bounce back signals from a transmitter on earth. (It is difficult to remember how new the sophisticated modern satellite communication system really is.) Arno Penzias and Robert Wilson, two scientists at Bell Telephone Laboratories in Holmdel, New Jersey, had been working on modifying a large, horn-shaped apparatus that had been used as a receiver in the series of experiments needed to test the ability of the satellite to return signals to earth. Because the satellite was in constant motion, this receiver could be pointed in any direction, despite its large size. This was very useful, because an ordinary radio receiver or radar antenna will be swamped by radiation emitted from sources on the ground and in the earth's atmosphere. Unless it can be pointed toward the sky an extraterrestrial signal would never be heard.

Penzias and Wilson intended to use their unusual antenna to do a rather prosaic survey of radio sources that lie outside the plane of our own galaxy. They found that there was an inexplicably large level of "noise" in their readings, equivalent to static on your radio or snow on your television screen. They spent a great deal of time trying to get rid of this signal, because they believed it came from their instrument and not from the sky. There are many ways that the instrument could produce such an effect, some involving complicated electronics and some quite commonplace. For example, at some time during the experiment a pair of pigeons took up residence in the antenna horn, coating a part of the apparatus with what was referred to as a "white dielectric substance." This source of error could be eliminated, of course, but no matter how many such extraneous effects were removed, the unexpected signal persisted.

In 1964 the invisible network that exists among working scientists began to go into operation. A young Princeton theoretician, P. J. E. Peebles, gave a seminar on the Big Bang theory at Johns Hopkins University, presenting some of his own work on the problem of the microwave radiation background. Word of his results traveled to the Carnegie Institute in Pittsburgh, the Massachusetts Institute of Technology, and Penzias in New Jersey via a series of telephone calls on subjects completely unrelated to the experiment. Once the theorists and experimenters got together, of course, the true nature of the noise in the antenna became obvious. It was in fact not noise that was being detected but the 3°K radiation that is a necessary consequence of the Big Bang. There is a folktale among astrophysicists that upon realizing this fact, one of the experimenters commented, "Either we've seen a pile of pigeon— — — —or the creation of the universe." In any case, it was not long before it was established that the radiation being detected was not associated with the antenna or with any particular galaxy or star but was truly universal. In the years that followed the announcement of the Bell Lab result, many independent measurements of the microwave radiation were made, all corroborating the original conclusions. For their work

Drs. Robert W. Wilson (left) *and Arno A. Penzias with the Bell Telephone Laboratories horn reflector antenna, which was employed on Project Echo and Telstar experiments and to measure unexplained noise radiation in the galaxy.* Photo courtesy of Bell Laboratories.

Dr. Arno Penzias (right) *checks the inside of the antenna. After carefully measuring and accounting for all the known sources of noise radiation from the earth's atmosphere and the galaxy, as well as from the antenna and associated receiving equipment, Penzias and Wilson found a residual amount of noise radiation that they could not explain. After consultation with colleagues at Princeton, they proposed the theory that this radiation is a result of the Big Bang.* Photo courtesy of Bell Laboratories.

Penzias and Wilson were awarded the Nobel Prize in physics in 1978.

The chain of events leading up to the discovery of the 3°K radiation illustrates one of the most important yet least understood points about scientific research: the most important discoveries are often made in experiments that are looking for something entirely different. This is what makes it so difficult to plan basic research in any rational way. You can never be sure when a quite ordinary experiment is going to get you involved in unexpected breakthroughs in unrelated areas of science.

One other point should be made about the cosmic microwave background. We have said that one of the founding precepts of modern science is that there is no preferred frame of reference. But is not the frame in which the Big Bang occurred and in which the microwave radiation is precisely isotropic (see Glossary) just such a frame of reference? If we take the analogy of the expanding gas seriously, then an observer who is not moving with respect to the universal fireball will see the same radiation whether he looks to his left or his right. On the other hand, an observer moving with respect to the fireball will see the radiation blue-shifted when he looks forward and red-shifted when he looks backward. (You can easily convince yourself of this fact by going through the type of argument we went through for the Doppler effect). Doesn't this situation violate the principle of relativity?

No, it does not. It says that there is only one frame in which the cosmic background is truly uniform, and observers in any other frame will detect slight shifts. However, the two observers in our example will see the same laws of nature operating in their respective frames of reference. Indeed, it can be shown that the requirement that they see the same laws of physics demands that they see different patterns for the cosmic radiation. There is nothing in the principle of relativity that requires that every observer see the same thing; they are only required to see the same laws. That only someone in one frame of reference sees a uniform microwave background is no more surprising than the statement that there is only one frame of reference in which the surface of the earth appears to be stationary.

In passing, we should note that delicate observations with microwave receivers flown in aircraft high in the atmosphere have shown that the earth is actually moving through the background radiation at the rate of 600 km/sec in the direction of the constellation Leo.

Conclusion

The events of the past half century have shown us that the Milky Way is definitely just one of millions of galaxies in the universe and that all of the galaxies are undergoing a uniform motion that can best be characterized as an expansion.

If we perform the mental experiment of reversing the expansion, we conclude that it has been roughly 15 billion years since the universe began in an event called the Big Bang. The age of the universe determined in this way is not known to a high degree of accuracy, but it does not need to be known exactly for any arguments that will be important for us in this book.

The most important and dramatic evidence for the Big Bang picture of the beginning of the universe comes from the discovery of universal microwave radiation. This radiation is of the type that would be associated with a body at a temperature of 3° above absolute zero—just the temperature that the expanding radiation gas resulting from the Big Bang should have reached by this time. In a sense, the microwave experiments are hearing the reverberations of the explosion with which the universe began.

Given this picture of our past, the goal of science is clear. Using known physical laws, we must trace back through the universal expansion, coming as close as we can to the moment of creation itself. Perhaps the most astounding fact is that we are now in a position to describe the universe as it existed during most of the first second of existence. To do so, we will have to learn something about the behavior of matter at very high temperatures, as well as something about the way matter is put together. Why we need this knowledge will become apparent as we think about those steps in the evolution of the universe closest to our own time, a subject to which we now turn our attention.

Chapter 2

The Ages of the Universe

All the world's a stage,
And all the men and women merely players:
They have their exits and their entrances;
And one man in his time plays many parts,
His acts being seven ages.

WILLIAM SHAKESPEARE
As You Like It

Atoms

The most intriguing thing about the Big Bang theory is that it tells us that the universe was not always in the state in which we see it now. The stars and the galaxies seem ageless to us, and indeed they are, compared with a single human life or the span of recorded history. But even the "eternal" stars were born at some time in the past, and they will certainly die at some time in the future. The best way to think about the time scale in relation to the birth of the universe is to imagine the entire 15 billion years since the Big Bang compressed into a single year, which I shall call the galactic year. On this time scale, all of recorded human history comprises no more than 10 seconds. The largest stars—those that burn fuel most quickly—live less than an hour. The sun and the solar system were created in September, and the sun can be expected to go on in very much its present state for a few more galactic months. The human race has been around for an hour or two, the exact time depending on how the current debate among anthropologists about the earliest human remains turns out.

The galactic year provides a useful way to visualize the most recent stages of the evolution of the universe, since the time scale involved in these stages is very large. Let us begin our description of the history of the universe by moving backward in time ("Big Bangwards") from the present, reckoning time in terms of the galactic year. If the present is December 31 of that year, then as we move backward through winter, fall, and summer, we shall see very little change in our surroundings. We may notice, as we pass backward toward January that the heavier chemical elements are becoming somewhat less abundant. This is because all elements heavier than helium, including the iron in your blood and the calcium in your bones, were first made in stars that eventually died and returned the elements they had manufactured to the interstellar medium, where they mixed with material forming new stars. The sun and the solar system, having formed relatively recently, have more of these materials than older systems. Aside from this rather minor effect, however, the universe would look pretty much the same to us in January of the galactic year as it does in December. If we stopped our time travel at any point and looked around, we would see most of the matter would be concentrated into galaxies, and those galaxies would be receding from each other. Therefore, for almost the entire lifetime of the universe, no fundamental changes have taken place.

In fact, one of the surprising things we shall learn as we explore the history of our universe is how much it resembles the seven ages of man into which William Shakespeare divided human life. Periods in which the appearance of things seems to stay more or less constant are suddenly followed by short bursts of revolutionary change, in which the aspect of the universe alters drastically and irreversibly. All of these periods of change are related to one central fact: the further back in time we go, the more the temperature of the universe increases.

The reason for this is not hard to understand. As we move backward in time, the matter in the universe becomes more and more compressed. It is a fact that such compression produces heat, as anyone who has used a hand pump to inflate a tire knows. The continued compression of air by the piston in the pump eventually makes the cylinder warm to the touch. As the temperature of an object increases, the molecules it contains move faster and faster. Since the molecules in any material undergo frequent collisions with one another, we would expect these collisions to become more violent as the temperature is increased.

If we imagine moving backward in time through the galactic year, we would see the galaxies (in very much their present form) getting closer and closer together. The effect would be like watching a movie run in reverse. The film would show the same thing all the way from the present (December 31) to within about an hour of the beginning of the galactic year on January 1. It would not be until we had reached the first hour of the galactic year (less than a million years in real time from the Big Bang) that any change in the universe would become evident. Still watching the film run backward, we would see the galaxies come together into a single undifferentiated mass of atoms. This would be a change, of course, but it would not be fundamental because even though the overall structure of the universe had changed, the basic unit of matter would still be the atom—an electrically neutral object in which as many negatively charged electrons circle the nucleus as there are positively charged protons within the nucleus itself.

Continuing to press backward in time, we would see this mass of atoms contract and heat up for about half an hour. At this point (about 500,000 years after the Big Bang in real time), the temperature would be high enough, and collisions between atoms violent enough, that electrons would be torn loose from their nuclei. Atoms would then cease to exist, and matter would appear in a fundamentally new state, one in which particles of opposite charge are free to move around independently of each other. This state of matter is called a plasma. Plasmas are routinely produced in physics laboratories these days, so we know a good deal about the way matter behaves in this state.

The transition from atoms to plasma can be thought of as somewhat analogous to the melting of an ice cube. At low temperatures, molecules of water are arranged in the rigid crystalline structure we call ice. As the temperature rises, it eventually reaches 32°F (0°C), at which point the motion of the molecules in the ice becomes so violent that it overcomes the forces that hold the crystal in shape. At that point the ice becomes water. A new form of matter (liquid) replaces the old (solid).

The primary difference between melting ice and the transition from a universe of atoms to a universe composed of plasma is that the plasma state does not occur at a single well-defined temperature. If a gas is heated, a gradual change occurs as more and more of its constituent atoms undergo violent collisions and have their electrons stripped off. Eventually, the temperature gets high enough so that any electron that happens to attach itself to a nucleus is quickly dislodged. In general, a relatively low temperature (a few thousand degrees) is enough to produce a fairly good plasma, and at a temperature of 10 million degrees (about one-tenth the temperature at the center of the sun), there will be no bound electrons in the material at all. The universe was at the higher of these temperatures a few hundred thousand years after the Big Bang and had cooled off to the lower before it was a million years old. The conversion from plasma to atoms was complete by the time the galactic year was 30 minutes old; all the rest of the year down to the present—virtually the entire lifetime of the universe—is taken up by the present era, the last stage in the development of the Big Bang. Although this is by far the longest era in the galactic year, it is also the least interesting period for scientists. It is not that nothing happens during the final stage, because this is when stars form, planets appear, and life develops. It is just that we know from firsthand experience what laws of nature are operating to bring these things about, and although a lot of important questions remain to be answered about the processes involved, someone who wants to think about fundamental questions naturally turns his attention to the first half hour of the galactic year. It is there, in the very beginning of the universe, that new knowledge waits to be discovered.

The present stage of the universe, therefore, was ushered in by the "freezing" of the hot plasma into a collection of atoms less than a million years after the Big Bang, almost 15 billion years ago. For most matter this new atomic state was transitory, because soon after this freezing occurred, another natural process began. Under the influence of gravity, the expanding material began to come together in clumps. These aggregations would eventually form the galaxies. Within each clump the gravitational forces continued to operate, drawing large clouds of gases together into nebulae. Even as the overall motion was universal expansion, matter was being drawn together on the local scale by the attractive force of gravity. When enough matter accumulated at a point, its combined gravitational pull attracted other matter in the vicinity, and the addition of the new material increased the gravitational force and brought in still more. As the mass of the body (by now a protostar) increased, the material was compressed and heated in a sort of reverse replay of the expansion following the Big Bang, although this time the drama was played out on a local, rather than universal, scale. Long before the temperatures at the center of the body reached the point where the fusion reaction could ignite to create a true star, atomic collision had again become sufficiently violent to strip the atoms of their newly acquired electrons and recreate the plasma state. In a sense, that process of gravitational collapse and atomic destruction undid the process of atomic formation that had taken place in the transition to the final stage of the Big Bang. As a result, most atoms possessed electrons for only a short period of time, perhaps less than a million years, between the moment at which the Big Bang had cooled off to where atoms could survive collisions and the time when those atoms were incorporated into the still forming stars.

In the next chapter, we shall discuss in detail one of the most striking and long-standing problems associated with the Big Bang cosmology. We shall see that the formation of galaxies and stars, with the subsequent return of much of the matter of the universe to the plasma state, could not have occurred until the initial explosion had cooled enough to allow atoms to form in the first place. Once this had happened, it is relatively easy to calculate how long this kind of accretion and clumping of matter would have taken. Until quite recently, there has been no generally accepted explanation of how galaxies could have formed quickly enough to account for the universe we observe today. It appears that by the time the atoms coalesced out of the plasma, matter was already spread too thinly for the galactic aggregation process to occur as described. The major transition that occurred approximately 500,000 years after the Big Bang, while it is easy to describe and visualize, is more complicated than it may appear to be on the surface. We will return to this point several times in the course of this book.

Nuclei

The stage in the life of the universe that ended with the formation of atoms 500,000 years after the Big Bang had its genesis in events that took place only 3 or 4 minutes after the moment of creation. The pattern of sudden change followed by a long period of relative quiescence, which we noted before, is seen here as well. Between the ages of 3 minutes and 500,000 years, the universe consisted of an expanding plasma with no atoms present. The nuclei in the plasma were protons (the nucleus of an ordinary hydrogen atom), deuterons (a type of hydrogen nucleus made up of one proton and one neutron), helium 3 (two protons and one neutron, usually written ^{3}He), and helium 4 (two protons and two neutrons, usually written ^{4}He). All other nuclei, as mentioned earlier, were synthesized in stars after the formation of galaxies.

Until roughly 3 minutes after the Big Bang, the temperature was so high, and the collisions in the plasma so violent, that no nucleus could cohere. Any nucleus that formed prior to this time would have been dismembered in these collisions. Thus, just as atoms could not exist until 500,000 years had passed from the moment of creation, the nuclei of atoms could not exist until 3 minutes had passed. The physical reason for both of these situations is similar: until the times indicated, the temperature of the universe was so high that collisions could break the bonds that would otherwise have bound their respective constituent particles together.

The temperature at 3 minutes was about a billion degrees—a little less than a hundred times hotter than the temperature at the center of the sun. But just as the transition from plasma to atoms did not occur suddenly but extended over a period of time, the earlier transition from protons and neutrons to nuclei was somewhat complicated. In addition to the reasons for the diffuse nature of the plasma-atom conversion discussed in the previous section, there is an additional difficulty associated with the fact that the process of nucleosynthesis, the formation of nuclei, involves a series of steps in a special sequence.

To understand what this means, think of a collection of protons and neutrons at high temperature. A proton and a neutron can collide and form a deuteron. If this nucleus lives long enough to encounter another proton, a nucleus of ^{3}He could be formed, followed by ^{4}He after another collision. Unfortunately, the deuteron is the weak link in this chain. It is the most weakly bound of all of these nuclei and therefore the one most likely to be broken up in a collision with any of the particles or the radiation in the plasma. Thus, the first step in the chain, the formation of a stable deuteron, will not occur until the temperature is well below the point where both ^{3}He and ^{4}He are stable. Once the temperature dropped below this level, at a time somewhat less than 4 minutes after creation, the rest of the steps proceeded quickly, and soon there existed the mixture of nuclei that characterized the universe for the next 500,000 years.

Elementary Particles

We can represent the evolution of the universe from the 3-minute mark to the present by the sort of process sketched in Figure 3. There are long periods when both the type of matter present and the physical processes going on are largely unchanged. From our present vantage point we can look back over a plateau dominated by the present universal expansion. The farther back we look, the hotter the universe appears, but that is the only change we note until our view encounters the abrupt step at about 500,000 years. Surmounting this step, we find another plateau stretching before us, corresponding to an increasingly dense and hot plasma. This terminates with another abrupt step around 3 or 4 minutes, when nucleosynthesis takes place.

If we extend our survey of the territory beyond the 3-minute mark, the situation becomes a little less simple. There are many processes that can go on at very high temperatures, each with its own characteristic starting temperature and each with its own characteristic effect on the universe as a whole. We can visualize this situation by saying that as we proceed backward past the 3-minute mark, we no longer encounter a plateau but instead see a series of hills and valleys as shown on the right in Figure 3. When a particular rise stops being a bump in the plateau and becomes the transitional step to a new era is largely a matter of semantics. For our purposes, we shall reserve terms like *stage, era, transition,* and *abrupt change* for those events in which either matter appears in a fundamentally new form or the forces that govern the behavior of matter appear in a fundamentally new form.

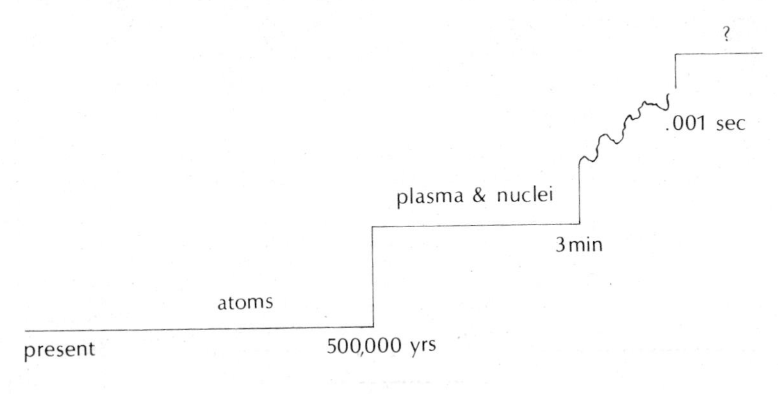

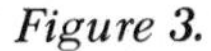
Figure 3.

Thus, the formation of atoms marks the beginning of a new era for us, while the formation of galaxies does not.

In these terms, the next true transition will occur before .001 second, or 1 millisecond, after the Big Bang. The time between .001 second and 3 minutes is far from uneventful, however. It represents the period when the universe is dominated by the interactions of elementary particles.* Some of the processes that make their appearance during this era are important examples of the things we will be seeing as we get closer to the beginning, so we will give a brief account of them here.

In the final analysis, all of the important events during the particle era depend on the interchangeability of mass and energy implicit in Einstein's famous formula $E = mc^2$. This formula tells us that if we have enough energy, we can create matter from it and, conversely, that if we have matter, it is possible to convert it into energy. Both aspects of this equation are verified daily at high-technology installations, by the conversion of energy into particulate matter in accelerators and the conversion of mass into energy in reactors. Indeed, the electricity used to generate the light you are using to read this book may very well have originated in the conversion of matter into energy in a nearby nuclear reactor.

NEUTRON DECAY

A neutron sitting alone in space will, without any outside interference, disintegrate. In place of the original neutron we will see a proton, an electron, and a massless, electrically neutral particle called the neutrino. We say that the neutron "decays" into these other particles, and write the reaction symbolically as

$$n \rightarrow p + e + \nu$$

where the Greek letter ν (nu) represents the neutrino.

What makes this decay possible is the fact that the mass of the neutron is slightly more than the combined masses of the proton and electron. The excess mass is converted into energy, energy that is associated with the motion of the three final decay products. Because the old term for electron produced in particle

*An excellent detailed discussion of this period of the Big Bang is given in Steven Weinberg's *The First Three Minutes* (New York: Basic Books, 1977).

and nuclear reactions is *beta ray,* this process is sometimes called the beta decay of the neutron. It is also possible for this sort of reaction to go in reverse—for energetic neutrinos to strike a proton, producing a neutron and a positively charged particle, or for such a neutrino to strike a neutron, producing a proton and an electron. All that is needed is for the neutrino to have enough energy to produce the additional masses of the final particles.*

In the early stages of the particle era, before the end of the first second, the temperature was high enough and collisions energetic enough for the neutrinos to convert protons to neutrons and neutrons to protons with equal facility. As the temperature fell, however, the energy needed to turn protons into neutrons was no longer available, so the former began to become more abundant. As the era progressed, the neutrons began to decay, which further reduced their abundance relative to the stable protons. By the time 3 minutes passed, there were only about one-sixth as many neutrons as protons, and by the time a half hour passed, all neutrons had either decayed or been bound into nuclei.

PARTICLE ANNIHILATION

In the next section we will see that for every particle in the universe it is possible to produce an antiparticle. The particle and antiparticle have equal mass, but in all other features are the opposite of each other. The antiparticle of the electron is called the positron. It has the same mass as the electron but a positive electrical charge. If an electron and a positron encounter each other, they undergo a process called annihilation. The particles disappear in a burst of energy, which in the case of the electron and positron takes the form of intense radiation.

It will be useful to think of radiation in the particle era (such as X rays or gamma rays) as being composed of particlelike objects called photons. One of the features of modern physics is that it is no longer necessary to think that there is a sharp logical distinction between particles like the electron and the waves that are usually thought to comprise radiation. Both particles and radiation display similar properties at the subatomic level, and the

*Experts will notice that I am using the term *neutrino* loosely here to refer to both neutrinos and antineutrinos. The distinction is not important for this discussion.

question of whether we should picture them as clumps of matter or waves becomes largely a matter of convenience. If we think of radiation as being composed of particles, electron-positron annihilation looks like Figure 4, in which the particles disappear and two photons take their place. The mass of the particles has been converted into the energy of the photons. The inverse process, in which a photon strikes some sort of target and produces an electron-positron pair (Figure 5), is also possible if the photon is sufficiently energetic. Consequently, we can think of the two processes pictured in Figures 4 and 5 as producing a sort of dynamical equilibrium in the universe. Every time a pair disappears through the process of annihilation, another pair will be created by an energetic photon elsewhere. After the 10-second mark, however, there is no longer enough energy available to replace the pairs that annihilate, and electrons and positrons begin to disappear in pairs. The energy released by this annihilation produces a momentary aberration in the Big Bang: it causes a slight rise in the temperature even while the expansion proceeds. The annihilation continues until all of the antimatter is gone and the present universe, in which only electrons exist in any abundance, emerges. The question of why there should be more matter than antimatter is a very difficult one and will be addressed in the next chapter.

PARTICLE CREATION

Up to now we have talked only of particles that are more or less familiar—protons, neutrons, and electrons. If we start to push back into the times before .01 second, however, we encounter a much stranger world. In this period the energy of the photons is high enough so that particle-creation processes of the type dis-

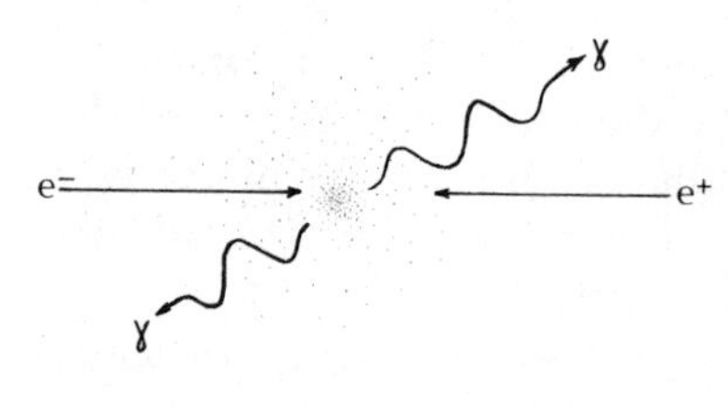

Figure 4.

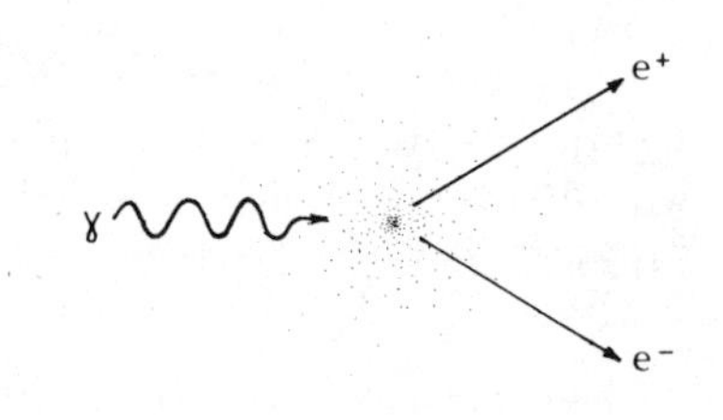

Figure 5.

cussed for electrons and positrons can also yield many other types of particles. To take a familiar example, there will be reactions in which protons and antiprotons will be produced in the interactions involving photons. All of the hundreds of elementary particles that have been discovered in the past few decades will be created. All such particles are unstable, and all decay much more quickly than the neutron. Consequently, for each species there will be an equilibrium reached between the rate of production in collisions and the rate of decay. High-energy theorists who have the inclination to do so can (and do) amuse themselves by working out the detailed demographics of the universe during this phase of expansion. For our purposes, we simply note that although many new kinds of particles show up at this time, none are fundamentally different from the more ordinary particles we have encountered. Thus, their appearance does not constitute a new era in the sense that we are using the term, but simply a proliferation of already existing types. In any case, after .01 seconds, all of these particles decay rapidly into various combinations of protons, neutrons, electrons, positrons, photons, and neutrinos.

The Rules of the Game

In the late 1970s, the particle era was as far back as it was possible to see. The reason why this should be so is not hard to discover. We possess the technical ability to produce in our laboratories all the phenomena we have encountered thus far. Electrons are stripped from atoms in ordinary electrical sparks, and large-scale plasmas are routinely studied in laboratories devoted to the development of fusion reactors. Since the nineteenth century we have observed the disintegration of nuclei in nature, and we have possessed the ability to take nuclei apart and put them back together again since the 1930s. The rapid development of particle accelerators since the 1950s has allowed us to produce elementary particles and to study them at will. This study has been one of the main areas of concern in modern physics.

Because of this sort of technology, we do not have to guess what will happen when a proton moving at a particular speed strikes a deuteron during the brief period of nucleosynthesis at 3 minutes. We can produce a proton of that speed and let it hit a deuteron in our laboratory and see what happens. The same remark holds for the behavior of elementary particles. All of the properties of matter and radiation in the universe from the .001-second mark to the present can be discovered (at least in principle) in the laboratory.

From .001 second back to the moment of creation, however, we no longer have this advantage. The temperatures involved are so high, and the particle energies so large, that we cannot reproduce them in the laboratory, and it is quite likely that we will not be able to do so within the lifetime of anyone reading this book. Consequently, .001-second marks a point of major change in our thinking about the processes following the Big Bang. From now on, we will be exploring a world so strange and unfamiliar that our only guide will be our theoretical ideas about the behavior of matter and energy.

This does not mean that we have to give up the idea that theories must withstand experimental tests before being accepted. This is, after all, one of the central tenets of modern science. What we will find is that the theories we develop will give us a very definite picture of the behavior of matter and energy at high energies and will also allow us to extrapolate and make predictions about what matter will be like at energies that are accessible to us. In this way, we can test the theories without having to reproduce the actual conditions of the early universe in our laboratories.

Despite the fact that we have to push into regions of energy and temperature that we have not, and probably cannot, explore experimentally, we can still do so in a way that is in harmony with the traditional way science is done and that does not simply introduce special ad hoc assumptions to explain what we find. So important is this aspect of modern cosmology in this age of creation "science" that we will state it as a basic rule (Rule I).

Rule I: The laws of nature that have operated at any time since the Big Bang still operate today and can be understood by theories which can be tested experimentally.

The philosophically inclined reader will recognize this rule as a statement of the doctrine of uniformitarianism, which first arose during the debates on geological evolution during the nineteenth century. It is not a statement that can be proved in the way that a theorem in geometry can be proved, but it reflects an important frame of mind among scientists. It is always possible to "explain" any known fact by tailoring a theory to fit it. Such explanations abound among believers in UFOs and other paranormal phenomena. They have the same validity in physics as Kipling's *Just So Stories* do in biology. If conventional theories simply cannot explain a given phenomenon, of course, unconventional ideas may become necessary. Until that time we will abide by Rule I.

Just as we see no reason to suppose that unusual processes were going on in the early universe, there is no reason to postulate special starting conditions for the Big Bang. For example, we saw that during the particle era electrons and positrons annihilated each other until only electrons were left. One way to explain this would be to assume that there were more electrons than positrons at the moment of creation. But this is no explanation at all—it merely assumes what we want to prove. Therefore, rather than make such arbitrary assumptions, we will assume that equal numbers of positrons and electrons were present at creation and look to the laws of physics to tell us how there came to be more of one than the other at a later time. Thus we come to our second rule.

Rule II:	No special conditions may be postulated at the creation.

This rule will play an important role in our discussion of problems with the Big Bang theory in the next chapter.

The rest of this book will be devoted to showing how we can use recently developed theories of elementary particles to trace the Big Bang back to a time within 10^{-43} second of creation. This development will take place entirely within the context of our two rules.

Previews of Coming Attractions

We can get some idea of what we will find when we move back from the particle era by thinking about what we have discovered so far. In each transition we had to understand two things, the structure of the dominant forms of matter involved and the forces that held these forms together. It would be impossible to understand the transition at 500,000 years, for example, unless we knew that atoms were made up of electrons and nuclei. Similarly, unless we understand the electrical force that keeps the electrons in orbit, we cannot tell how high the temperature would have to climb before collisions will tear the electrons loose. Complete understanding of any further transitions in the early universe, then, requires that we study the structure of the elementary particles and the fundamental forces of nature. A description of our new theoretical understanding of these subjects will be the topic of the second section of this book.

In the meantime, to arouse the reader's curiosity and give him or her the courage to cross the frontiers of particle physics, the following table is a summation of what has been covered so far and a glimpse of what is to come. Some of the terms may seem strange or unfamiliar right now, but by the time you have finished this book, this table will be clear and understandable to you.

Age of Universe	*Transition*	*Name of Era Following Transition*
500,000 years	Plasma to atoms	Present (atoms)
3 minutes	Formation of nuclei	Plasma and nuclei
.001 second (1 millisecond)	Formation of particles from quarks	Particle
10^{-10} second	Weak and electromagnetic forces unify	Quark
10^{-35} second	Electroweak and strong forces unify	Electroweak
10^{-43} second	All forces unify	GUT
?	Quarks and bosons become interchangeable	Ultimate simplicity, supersymmetry
0	Vacuum to matter (?)	Big Bang

One point of terminology concerning the table ought to be cleared up before we proceed. I have been using the term *millisecond* somewhat loosely to refer to the point at which the particle era begins. In point of fact, the exact time at which this occurs is a subject of some debate, as we shall see in Chapter 9. We know that the particle era had not begun when the universe was a microsecond (10^{-6} second) old and was well established by the time it was a millisecond old. The best current estimates seem to put the beginning somewhere between .01 and .001 millisecond (*i.e.*, between 10^{-4} and 10^{-5} second). Rather than bring up this uncertainty every time I want to talk about the onset of the particle era, I will use the terms *millisecond* or *first millisecond* rather loosely to refer to the somewhat ill-defined time at which this process occurred in the early universe.

Chapter
3

Problems with the Big Bang Theory

Roses have thorns, and silver fountains mud;...
And loathsome canker lives in sweetest bud.

WILLIAM SHAKESPEARE
"Sonnet XXXV"

No scientific theory is perfect. In each there is some small area of uncertainty—some problem that has not been solved. If it were not so, it would scarcely be worthwhile doing science at all. The unsolved problems are the focus of research and often acquire an importance for individual scientists far beyond any intrinsic interest they may have. The Big Bang theory is no exception to this general rule. It, too, has had difficulties in explaining certain aspects of the present state of the universe in terms of the basic rules laid down in Chapter 2. Four of these problems could be considered to be fundamental, in the sense that they involve questions about the basic processes involved in the evolution of the universe.

The Antimatter Problem

In 1932, Carl Anderson, then a junior faculty member at the California Institute of Technology, discovered a new kind of particle in an apparatus he was using to study cosmic rays. Christened the positron, this particle has the same mass as the electron, but instead of a negative electrical charge, it has a positive charge. It was the first example of antimatter to be seen in the laboratory. For this discovery, Anderson was awarded the Nobel Prize in physics in 1936, becoming the only man to receive the Nobel Prize before he was awarded tenure at his university.

We have already mentioned some of the more important properties of antimatter. We know that it undergoes annihilation when it encounters ordinary matter and that it can be created in energetic interactions between elementary particles. If we have a collection of particles and antiparticles at a very high temperature, we would expect a balance to be set up between these two processes: every time a pair annihilated each other at one point, another pair would be created in an unrelated collision elsewhere. As the temperature falls, however, it must eventually reach a point where the creation process can no longer proceed, since there is not enough energy to produce the mass of the particle pair. At this point, the balance disappears and annihilation proceeds alone. Each time a particle meets an antiparticle, the pair disappears and is not replaced. This goes on until there is no longer any possibility of annihilation because all of the particles (or all of the antiparticles) are used up.

Take an electron-positron mixture as an example. Suppose that at the time when the temperature falls below the point where creation is possible, there are 1,000 electrons and 500 positrons. Random encounters between electrons and positrons would take place as a matter of course, and each encounter would result in one less pair in the system. Eventually, all 500 positrons, with their accompanying electrons, would be converted into photons, and the system would contain only the 500 remaining electrons and radiation. No further annihilation is possible at this point, for the simple reason that there are no positrons left to be annihilated. Thus, the eventual makeup of the system will depend on the net particle number (the number of particles minus the number of antiparticles) at the moment when pair creation stops.

Since for every particle there is a corresponding antiparticle, both the annihilation and creation processes we have outlined for the electron-positron pair can occur for other types of particles as well. For example, there is an antiproton, which has the same mass as the proton but a negative (instead of a positive) electrical charge. Proton-antiproton pairs can be created by high-energy photons, and when a proton encounters an antiproton, the pair annihilates. Since the mass of the proton is 1,836 times that of the electron, however, the temperature of the universe has to be a great deal higher to allow the creation of proton-antiproton pairs than it does to allow the corresponding process for electron-positron pairs. Therefore, as the universe cools off, the temperature will fall below the point at which the proton-antiproton pair can be created long before it reaches the corresponding point for electrons. Thus, there will be a sequence of unbalanced annihilations, one corresponding to each type of particle-antiparticle pair that can be produced. The last great annihilation scenario, the one involving the electrons and positrons, took place about 13 minutes (in real time) after the Big Bang.

However, these details about the behavior of matter and antimatter in the early universe are not nearly so important as one crucial fact: after the beginning of the particle era, there is no known process that can change the net particle number of the universe. Both annihilation and creation involve pairs of particles, and for every particle that is created or destroyed, the corresponding process occurs for an antiparticle as well. Thus, by the time the universe is a millisecond old, the balance between matter and antimatter is fixed forever. All that can happen after that time is for the drama of annihilation to play itself out as we have outlined.

One of the striking facts about the earthly environment is the noticeable scarcity of antimatter. Small amounts can be created in specialized laboratories, but all of the antimatter created in the history of science would not fill a thimble. Our satellites and planetary probes have landed on, or passed near, most other important bodies in our solar system and brought back the same verdict: no antimatter anywhere. Indeed, the fact that the Viking landers did not explode when they touched the surface of Mars is proof positive that Mars, like Earth, is made of matter.

Although we cannot yet send probes to great distances outside of our solar system, we can nonetheless sample the material elsewhere in our galaxy by examining cosmic rays. These are particles generated in distant stars, accelerated, and sent out through space. We can identify them when they happen to enter a counter in some laboratory on earth. Most cosmic rays are protons or the nuclei of light atoms. The slow-moving ones originate in the sun, but the more energetic variety comes from other parts of the galaxy. A small fraction of the cosmic rays are moving so fast that they could escape from our own galaxy, and we assume that at least some of these extremely energetic particles must have escaped from distant galaxies and traveled to our own.

Therefore, even though we are temporarily locked into our own neighborhood of space, we constantly receive visitors from elsewhere. By studying these visitors, we can learn something of the region from which they came. As far as the matter-antimatter question is concerned, such studies yield an unambiguous result: our galaxy, and most likely all of the galaxies in our region, are composed of matter. The earth is quite typical, therefore, of a large body of space. Only matter exists here, with a tiny amount of antimatter created in natural or man-made collision processes. The question of how to explain this apparent imbalance in nature is known as the antimatter problem.

The antimatter problem can be resolved in only two ways: either there was a preponderance of matter over antimatter when the Big Bang entered the particle era, or the antimatter in the galaxy has somehow segregated itself from the matter, and some of the more distant galaxies are, in fact, antigalaxies. If indeed there was an imbalance at the one-millisecond point, then there are two ways in which it could have arisen: either the universe started out with more matter than antimatter, or there is some process in the period before the start of the particle era that produced more matter than antimatter. Let us examine these possibilities.

ANTIMATTER HAS BECOME SEGREGATED

This idea, first proposed by Swedish Nobel laureate Hannes Alfvén, was very popular during the 1960s and early 1970s. Its most popular version went like this: At the start of the particle era, the universe was indeed composed of equal parts of matter and antimatter, but the mixture was not uniform. It was more like a tossed salad than a milkshake, with some bits containing more matter than antimatter and some containing just the opposite. In each separate region, the annihilation process operated to eliminate the minority particles, resulting in a region that was either all matter or all antimatter. At the boundaries between different regions, a well-known physical process occurred. The particles and antiparticles that encountered each other at the boundaries between regions annihilated, producing a flood of gamma and X rays. This radiation literally blew the remaining material away from the boundary region, preventing the annihilation from going more quickly. A similar phenomenon occurs when drops of water skitter around on a hot skillet. In this case, the evaporation of the water on the droplet's underside creates a layer of steam, called a Leidenfrost layer, which insulates the rest of the droplet from the pan. Thus, the droplet lasts much longer than you might expect. A similar situation occurs when an Indian fakir walks over glowing coals, only in this case it is perspiration that creates the Leidenfrost layer. In the early universe, a Leidenfrost layer could be expected to form between regions of matter and regions of antimatter, delaying the ultimate annihilation of the two regions.

Until recently, this hypothesis could not be refuted because the gamma rays emitted in the annihilation process are absorbed in the earth's atmosphere and cannot be detected by telescopes on the earth's surface. With the advent of rocket and satellite X-ray astronomy, however, this limitation was surmounted. Observations made above the atmosphere have indeed identified many sources of X rays in the sky, but none of the sources corresponds to the kind of extended boundaries you might expect from a Leidenfrost layer. From this we conclude that our own galaxy and our own cluster of galaxies are composed entirely of matter. We cannot, of course, rule out the possibility that other clusters of galaxies may be antimatter, but such a situation would require that the initial clumping of matter and antimatter be very large; it would be like tossing a salad and finding all the tomatoes on one side of the bowl and all of the lettuce on the other. While we cannot prove that there are no regions of antimatter anywhere in the universe, there is no evidence whatsoever to suggest that there are. In the face of this result, interest in this particular solution to the antimatter problem has waned considerably in recent years.

THE BIG BANG STARTED WITH MORE MATTER THAN ANTIMATTER

This solution cannot be ruled out either, since it is impossible to go backward in time and see the Big Bang. Perhaps, if no other solution can be found, we would be forced to fall back on this one, but it does constitute a violation of Rule 2. Besides, there is an inherent ugliness in assuming what should be proved, for it simply leads us to the obvious next question: Why should the universe have started this way? To a physicist, the only net particle number that does not need to be explained is zero.

SOME PROCESS PREVIOUS
TO THE PARTICLE ERA IS RESPONSIBLE

Advances in astronomy and our own philosophical biases leave this possibility as the only way we can approach the antimatter problem. That means that if we expect to clear up this particular difficulty, we are forced to explore the stages through which the universe passed before it was a millisecond old. When we do so, we encounter one important roadblock. Since the discovery of antimatter, physicists have believed that the laws of nature are almost completely symmetrical as far as the distinction between particles and antiparticles is concerned. We will have to explain, therefore, how such a set of laws operating on a universe whose net particle number is zero could produce a universe in which matter predominates and antimatter is virtually nonexistent.

The Galaxy Formation Problem

We have already alluded to the fact that the process by which galaxies formed is intimately connected to the formation of atoms about 500,000 years after the Big Bang. Prior to this time, the matter in the universe was in the form of a plasma, with free electrons and nuclei moving about independently of each other. Interspersed with this collection of charged particles was a very hot sea of radiation in the form of photons.

It turns out that when a photon encounters a free electrical charge like the electron, a very dramatic interaction takes place. The photon bounces off the particle, a process physicists call scattering. You can think of the process as being analogous to the collision of two billiard balls at high speed. Both bounce away from the interaction point in a direction different from the ones at which they approached it.

The continuous collisions between photons and particles in the early plasma created a pressure that prevented matter from collecting into galaxy-sized chunks. The best way to picture this pressure is to think of an ordinary automobile tire. The air molecules inside the tire collide with the tire and bounce back. Each such collision results in a small force being exerted on the rubber, and the sum of the forces exerted by all of the molecules, what we call the tire pressure, keeps the tire inflated. The crucial point is that the pressure exists because the molecule bounces back when it meets the surface of the tire. In the language we are using, a tire stays inflated because the rubber walls are very efficient at scattering the air molecules.

It is clear that if we replaced the rubber tire by something that was not an efficient scatterer—cheesecloth, for example—no pressure would be exerted on the wall. Molecules coming to the boundary would pass through unscattered, and the tire would collapse. Now, the atoms into which the universe froze at 500,000 years are rather inefficient scatterers of radiation. You can convince yourself of this fact by remembering that it is not at all unusual to be able to see a hundred miles on a clear day. (I can remember, for example, standing in Los Alamos on a typical New Mexico night and seeing the street lights of Albuquerque ninety miles away.) This means that light can travel long distances through the atoms that comprise our atmosphere without undergoing tremendous distortion.

As far as the early universe is concerned, as soon as atoms formed the radiation was no longer strongly scattered by matter. This had two immediate consequences: (1) the radiation was free to expand without interference (the microwave background is the result of this expansion), and (2) the radiation no longer exerted pressure on the matter. After 500,000 years, then, it was possible for matter to coalesce into the galaxies that dominate the universe today.

The process by which this could occur is easy to describe. Suppose that the matter in the universe was spread out uniformly after 500,000 years. Normal atomic motion will guarantee that sooner or later two or more atoms will find themselves closer together than normal. When this happens, the two will exert a somewhat stronger than normal gravitational force on neighboring atoms, pulling these new atoms into the collection. The result is that still more atoms will be collected in the original spot, exerting a still stronger gravitational attraction on their neighbors. It is clear that this process will go on with more and more material aggregating around the original point, until all of the available material is pulled in. We would expect that even a perfectly smooth universe would eventually segregate itself into discrete collections of matter, and it is these collections we identify with the galaxies.

The problem with this simple picture of galaxy formation is not that the process we have described will not occur but that it takes too long for a collection of a few atoms to build up to a galactic mass. The important point is that the aggregation process is taking place in a universe that is undergoing a rapid expansion. If the collections of matter do not reach a critical size quickly, the universal expansion will have carried the rest of the material out of their reach before enough mass has been accumulated to form a galaxy. Astronomers have known for quite a while that the time required for random atomic motion to produce galaxy-sized aggregates is quite long. If this were the only way the process could occur, there would be no galaxies in the sky at all. The essential problem is that the initial part of the process—building up from a few atoms to a sizable lump of material—is very slow. It appears, therefore, that the universe could not have entered the era of atoms with a smooth distribution of matter.

But this is not too surprising, for you would expect anything as chaotic as the Big Bang to produce a highly turbulent cloud of expanding gas. There would be swirls and eddies in the plasma, and these would correspond to ready-made concentrations of mass to start the accumulation process. This idea was quite popular for a while, but numerous calculations showed that even starting off with normal turbulence in the plasma, the gravitational attraction is still too slow for galaxies to form. Therefore, for almost a decade, the question of how the large-scale structure of the universe could have come into being has been a major unsolved problem in cosmology.

Since the simple solutions do not seem to work, there are only two ways out. Either the universe was created with matter already clumped together into aggregations that would eventually form galaxies, or there is some as yet unknown process that would form such aggregations before the formation of atoms. The first option is a direct violation of our rules and will therefore be avoided as long as possible. The second option seems more inviting, but we know that there is no process (other than turbulence) that could produce the necessary clumping of matter during either the particle or nuclear eras. As was the case with the antimatter problem, we are forced to look to the period before 1 millisecond to explain the existence of galaxies.

The Horizon Problem

Microwave background radiation plays an important role in our knowledge of the early structure of the universe, as we have already seen. There is one approach to this radiation that, although it seems on the surface to be eminently reasonable, leads to difficulty if it is pursued. When microwave detectors are flown high in the atmosphere in balloons or aircraft, the radiation is found to be isotropic to an accuracy of better than .01 percent. This means that if we measure the photons coming from one direction of the sky and then turn 180° and perform the same operation, we will find that the two batches of photons are identical to within that accuracy.

At first glance, this seems to be what we would expect on the basis of our intuition. Why should one portion of the sky be different from any other? But if you recall the origin of the 3°K radiation, you will remember that what we see now depends on the temperature of the region of space from which the radiation was emitted 500,000 years after the Big Bang. The problem is that radiation now reaching our detectors from opposite ends of the universe was emitted from sections of the freezing plasma that were at more than 500,000 light-years apart. If this is true, how could it have happened that the two regions were at exactly the same temperature?

To understand why a situation like this gives physicists gray hairs, let us consider a simple analogy. Suppose we had a bathtub filled with water in such a way that all the water to the right of a particular line was hot and all the water to the left was cold. We all know what would happen in this situation. If we waited long enough, all the water in the tub would eventually come to the same lukewarm temperature. In the jargon of physics, we say that the water in the tub would "come to thermal equilibrium."

How long would it take for this process to be completed? That would, of course, depend to a certain extent on the temperatures involved, but our experience tells us that in most cases a half hour would be sufficient. But suppose we ask another question. Suppose we want to know the shortest possible time, in principle, that the water could take to come to equilibrium.

We know that there is a correlation between the temperature of the water and the motion of molecules that constitute it: the higher the molecular speed, the higher the temperature. In the movement toward thermal equilibrium, the fast molecules in the hot water slow down by colliding with molecules from the cold water. Clearly, we could not establish thermal equilibrium in a time any shorter than the time required for a molecule to move from the hot to the cold region of the tub. The actual process of sharing out energy among the water molecules may actually take a good deal more time than this, but it certainly cannot take less. To use another bit of physics jargon, we say that the process of coming to thermal equilibrium cannot take place unless the two regions whose temperatures are being mediated can "communicate" with each other. In anthropomorphic terms, you can think of thermal equilibrium as being dependent on a messenger going from one region to the other and saying, "This is what the temperature is over there."

In a bathtub the communication between different regions is done through the motion of molecules. In the early universe it was done by both light and particles. We know from the theory of relativity that information cannot be transmitted from one region to another at a speed faster than the speed of light. Thus, when we observe the isotropy of the microwave radiation and conclude that it all was emitted from regions with a common temperature, we are in effect saying that the universe was in thermal equilibrium 500,000 years after the Big Bang. This implies that all parts of the universe must have been in communication with each other before that time.

And this brings us to a problem. If we go back to our analogy of the universe and the balloon, in which we envisioned the galaxies as dots on the balloon's surface, then the light going out from a given galaxy would be represented by a group of ants moving out in an ever-widening circle centered on that galaxy. All of the galaxies within that circle would have received light from the central galaxy and, conversely, could be seen by an observer in that galaxy. A galaxy that at any moment is still outside the circle of ants would not be visible to the observer, simply because light would not have had time to travel to him from that particular source. The outward moving circle of ants represents the observer's horizon—he can see anything within the horizon, but nothing outside it. Just when a particular point comes within our horizon depends on the speed of the ants' movement and the speed of the balloon's expansion.

If we now think about the emission of microwave radiation at 500,000 years, it turns out that radiation emitted from opposite portions of the sky comes from regions that were outside each other's horizons at that time. In other words, there was no way that these regions could have communicated with each other and established thermal equilibrium. How, then, could they have come to be at the same temperature to within .01 percent? This is known as the horizon problem.

What are the possible solutions? As with the other problems we have discussed, once we enter the era of particles we know of no way to find a resolution. If the universe is not in thermal equilibrium by that time, it will never be. Furthermore, the expansion of the universe during the first millisecond was so rapid that it can be argued that there was never a time when communication could have been established over the entire universe.

Therefore, we are left with two possibilities. Either the universe was created in thermal equilibrium, or some process before the era of particles precludes us from making the kind of naive extrapolation that leads to the paradox. The first possibility is equivalent to assuming that all the water put into the bathtub was lukewarm to begin with, an assumption that violates our rules. The alternative, once more, demands that we look more closely at what preceded the era of particles and hope that some new discoveries there will resolve the problem.

The Flatness Problem

Once we have the picture of the expanding universe firmly in mind, we can ask a very simple question: Will the expansion go on forever or will the expansion someday reverse itself, leading to a period of universal contraction? The answer to this question depends on the amount of mass in the universe. If there is enough mass, the gravitational force it exerts will be sufficient to slow down and eventually stop the receding galaxies. If there is not enough mass, this will never happen. An exactly analogous situation occurs when you throw a ball up from the surface of the earth. Because the earth is very massive and exerts a strong gravitational attraction, the ball will slow down, stop, reverse its motion, and fall back to the surface. The same ball, if launched from the surface of a small asteroid, would sail off into space because the mass of the asteroid would be insufficient to stop it.

The question of whether the universe will continue to expand or will someday contract—the question of whether it is "open" or "closed," to use a cosmologist's terms—would seem to be a simple one. All we have to do is add up all the mass in the universe, calculate the gravitational attraction on the receding galaxies, and see if it is enough to stop the recession. Simple in principle, this prescription is hard to carry out because all the matter in the universe is not necessarily visible to us. We will discuss this fact in great detail in Chapter 14, when we consider the ultimate fate of the universe. For the moment, however, we note that observations indicate that the amount of matter in the universe is surely greater than one-tenth and surely less than ten times the critical amount needed to stop the expansion.

The universe, in other words, is either barely open or barely closed. In the language of cosmology, we say that the universe is "very nearly flat," a flat universe being one in which there is just enough mass to bring the expansion to a stop. The problem is that given all of the infinite possible masses that the universe could have, why does it have a mass so close to this critical value? Why is the universe almost flat?

By now, you have probably realized that Rule 2 plays an important role in our thinking about the early universe. The distaste that scientists feel when they are forced to assume what remains to be proved is a powerful motivating force in research. But there is one aspect of the rule, a hidden assumption, that must be explored.

When we say that something "should" be proved or "ought to be" explainable without initial assumptions, we are, in fact, making a judgment about the state of scientific knowledge at the time the words are pronounced. We are disturbed by being driven to make special assumptions about the creation of the universe only if we feel that we know enough about the processes involved to explain the facts without any assumptions at all. For example, we have studied antimatter in our laboratories for more than half a century and feel that we know what there is to know about it. The existence of the antimatter problem flies in the face of this belief; we are therefore willing to expend a great deal of effort to solve it. The three problems we have talked about so far are all of this type. For the sake of clarity, let us call problems that are presently unsolved, but ought to be soluble with our present knowledge, problems of the first kind.

There are other kinds of problems in cosmology that do not fit this definition. For example, if I were to ask why the charge of the electron has the value it does and not some other value, no one could provide a satisfactory answer. This is a genuine unsolved problem, but scientists would not fault a theory for being unable to answer it. It is recognized that a question such as this requires a much deeper understanding of physics than we have at the present time and that it is unlikely that any theory we produce in the near future will have anything to say about the electron's charge. Thus, while we regard a failure to resolve the antimatter, galaxy formation, and horizon problems as a major indictment of any cosmological theory, we do not put the electron problem in the same category. Problems of this type, which are generally felt to require a much deeper understanding of the world than we now possess, we will call problems of the second kind.

The so-called flatness problem is actually a problem of the second kind that is traditionally lumped together with problems of the first kind in discussions of the Big Bang. It must be kept in mind, however, that it is qualitatively different from the other questions we have raised in this chapter. I can imagine a scientist giving up his belief in the Big Bang theory if the antimatter problem remained intractable. I cannot imagine him doing so because of a failure to resolve the flatness problem.

It would appear that the total mass of the universe would be one of those quantities, like the precise value of the electric charge, whose explanation must wait for a more fundamental theory than any we now have available. Nevertheless, if it should turn out that our exploration of the first millisecond of the life of the universe produces a solution to this problem, we shall have the right to regard that solution as an unexpected bonus and to use its existence as a strong argument in favor of our analysis.

Summary

There are four fundamental problems associated with our picture of the Big Bang. Three of these are problems of the first kind, and a failure to resolve them would have to be taken as evidence of a major weakness in our understanding. These problems are (1) why there is so little antimatter in the universe, (2) how the galaxies could have formed in the time allotted for this process, and (3) why the universe is isotropic. In addition, there is one problem of the second kind that is traditionally associated with the three problems of the first kind: why the mass of the universe is so close to the critical value required to close the universe.

We have seen that if we are to avoid making special assump-

tions about the moment of creation in order to solve these problems, we have to explore the period of time before the beginning of the particle era—the first millisecond of existence. In the second part of this book, we present a discussion of the theories that allow us to do so.

PART TWO

Tracing the first millisecond of existence requires some understanding of the processes that dominate the behavior of matter at very high energies. Chapters 4–8 describe our current understanding of this field, as incorporated in the so-called unified theories. Realizing that not every reader may wish to go through all of the detailed arguments involved in establishing these theories, we have provided a "fast track" through this part of the book. At the end of each chapter, in a different typeface, there is a summary of the main results of that chapter. The reader who is interested in getting on to the description of the early universe can read these summaries and move right ahead to Chapter 9, while the reader who wishes to acquire a deeper understanding of how the ideas in the summaries came into being can go through Chapters 4–8 in more detail.

Chapter
4

Elementary Particles and Quarks

Things are seldom what they seem.
W. S. GILBERT AND ARTHUR SULLIVAN
HMS Pinafore

The Concept of the Elementary Particle

There is a long tradition in Western thought which holds that in order to understand something fully, it is necessary to take it apart and see how it is made. The phrase *see what makes it tick,* with its implication of performing this sort of operation and finding the internal gears and wheels that make something go, is one familiar statement in this mode of thought. For almost two centuries, scientists trying to discover the nature of matter have been uncovering successive layers of "reality" in an attempt to find a truly elementary particle—a bit of matter that will explain how the world ticks. The goal of this endeavor is to explain the entire material world in terms of the different arrangements of a few basic entities. In the same way, one might try to explain architecture by discovering a few basic elements (bricks and beams, for example) and noting that all buildings, no matter how complex, are the result of different arrangements of these basic parts.

In the early nineteenth century, British chemist John Dalton proposed that the elementary building block of matter was the atom (literally, "that which cannot be divided"). The infinity of chemical substances that were coming to be known then were all simply different arrangements of a few different kinds of atoms. Early in the twentieth century our knowledge penetrated to a deeper level, when Ernest Rutherford showed that the atom, far from being indivisible, had a very definite structure in which negatively charged electrons circled a positively charged nucleus, much as the planets circle the sun. For a while, it appeared as if the final solution was in sight—a world in which everything was some combination of protons, neutrons, and electrons. The protons, with a positive charge, and the neutrons, electrically neutral, are put together in roughly equal numbers to make up nuclei, and the electrons are added in their orbits to make atoms.

Unfortunately, this sweet dream of simplicity was not to be. Researchers began to probe the nucleus by allowing very fast beams of protons to collide with it. These fast beams were either supplied by nature in the form of cosmic rays or by scientists employing the powerful particle accelerators that were being developed. In the debris of these collisions, new particles, neither protons, neutrons, nor electrons, were seen. As our ability to impart high energies to the projectiles in these experiments improved, it became possible to create more and more different kinds of particles in our laboratories by accelerating protons to ever higher velocities and then allowing the speeding particles to hit a stationary target. In the collision process, some of the energy of motion of the proton was converted into the mass of new species of particles. By the mid-1970s, a special 246-page edition of the professional journal *Reviews of Modern Physics* was published to provide a relatively compact summary of all of the data that had been gathered on the elementary particles discovered up to that point.

The story of how these particles came to be found and what the discoveries mean is too long to be told here.* For our exploration of the early universe, we will only need to have at our disposal some of the properties that these particles have in common and a rudimentary knowledge of the classification schemes that have been developed to deal with them.

Particle Properties

MASS

Every particle has a definite mass, although that mass can sometimes be zero. We know that mass and energy are interchange-

*It is available for the interested reader in my book *From Atoms to Quarks.* (Scribners, 1980).

able in the ratio given by the Einstein formula $E = mc^2$. Thus, if collisions between particles become vigorous enough, there can, in principle, be enough energy to create a particle of any desired mass. This creation of particle mass from energy is different from the kind of particle production encountered in some collisions with atoms and nuclei, where the energy of the projectile simply tears loose the already existing constituents of the target system. In creation processes, a particle is brought into existence where none existed before.

Even the unit that high-energy physicists use to measure the masses of particles depends on this mass-energy equivalence. Masses are customarily measured in gigaelectron volts (GeV). One electron volt (eV) is the energy required to move a single electron across 1 volt of electrical potential. An electron moved from one pole of an ordinary automobile battery to the other, therefore, acquires an energy of 12 eV. To move ten electrons would require 120 eV, and so on. The prefix *giga* represents what Americans call a billion and what Europeans call a thousand million and what both would call 10^9. A gigaelectron volt, therefore, represents the energy required to move a single electron across a 10^9-volt battery or a billion electrons across a 1-volt battery. It is a unit of energy, but because we know that there is an equivalence between mass and energy, it can serve as a unit of mass as well. For reference, the energy equivalent of the mass of the proton is slightly less than 1 GeV, and typical masses of particles seen in accelerators run from .005 GeV (the electron) to about 100 GeV (the vector boson, a particle we will encounter later). There are a few important particles whose mass is zero.

SPIN

You can visualize something about the properties of elementary particles by imagining them to be tiny billiard balls. One of the things a billiard ball can do is spin around on its axis, much as the earth does. It is this property of the particles that we call spin.

We are used to thinking that large bodies such as the earth can spin at any rate whatsoever and that the earth's twenty-four-hour period of rotation is the result of historical accident. In principle, the earth could have ended up rotating every twenty-three hours or even every year.

When we deal with particles at the quantum level, however, this type of freedom in the spin no longer exists. We find that an elementary particle can spin only at certain well-specified rates. To understand what quantum mechanics tells us about particle spin, we can introduce the concept of the basic rotation rate. This is a fundamental unit in terms of which the actual spin of the particle is measured. Quantum mechanics tells how to calculate this rate for any particle*. The proton, for example, has a basic rotation rate of about 10^{22} complete revolutions per second.

The significance of the basic rotation rate is this: particles can spin only at rates that are certain well-defined multiples of their basic rotation rate. They can, for example, spin twice as fast or half as fast, but they cannot spin three fourths or two thirds as fast. In the language of quantum mechanics, we say that the spin of the particles is quantized. In a sense, defining the basic rotation rate for any particle is like defining the basic unit of currency. Once we have defined the dollar, we can then go on to say that the denominations of coins are quantized. There are dimes and quarters, for example, but no coin whose value is thirty one cents.

An individual object such as the earth or the proton spins at a definite rate that is some multiple of its basic rotation rate. The shorthand way of designating that rate is a single spin number. For the earth, this multiple is a huge number (on the order of 10^{61}). For the proton the number is ½. In other words, we find experimentally that the proton spins at precisely half of the basic rotation rate that quantum mechanics assigns to it. In the jargon of high-energy physics, we say that the proton has *spin* ½ or is a *spin-½ particle.* Particles that rotate exactly at the basic rotation rate are said to have spin 1, particles that do not rotate at all have spin 0, and so on. Any particle we talk about will have some rotation rate, and this rotation rate is constant in time.

STABILITY

With a few exceptions which we will note explicitly, every elementary particle is unstable. By this we mean that if we watch the particle we will eventually see it disintegrate, and in its place

*Technically, the basic rotation rate is calculated by setting the angular momentum of the rotating particle equal to Planck's constant. For a sphere of mass M and radius R turning once every T seconds, this corresponds to the requirement that $4\pi MR^2/5T = h$. The value of T obtained from this equation is what we are calling the basic rotation rate.

we will see other, lighter particles. We call this process the decay of the original particle, and the particles that appear the decay products. For example, a free neutron will eventually disappear, and in its place will be a proton, an electron, and a massless particle called the neutrino. These decays can be relatively slow, as in the case of the neutron (which will last more than 10 minutes before disintegrating), or they can be very rapid indeed, with some particles undergoing decay in as short a time as 10^{-23} second. Nevertheless, the decay of massive particles is a crucial aspect of the subatomic world, and it provides us with a means of classifying the bewildering assortment of elementary objects found by the experimenters.

Classification Schemes for Elementary Particles

With hundreds of particles discovered, and almost all of them unstable, it is pointless to try to designate some as elementary and others as not elementary. Even the proton, that most "elementary" of particles, may decay if we wait long enough (a possibility discussed in detail in Chapter 8). Either all of the particles are elementary in the sense we are using the term or none of them are. The best we can do, therefore, is to find some way of organizing and classifying the particles, in the hope that doing so will bring some order to our thinking about them. To carry out this classification, we will use some of the properties of the particles we have just discussed. It should be kept in mind during this exercise that every classification we make of the particles has a sort of mirror image in the classification of the antiparticles—an image we will not elaborate, for the sake of brevity.

CLASSIFICATION BY SPIN

Particles that have half-integer spin—½, 3/2, 5/2, and so on, in the terminology introduced in the previous section—are called fermions, after the physicist Enrico Fermi, who first elucidated their properties. This category includes the proton, electron, neutron, and a good number of other massive particles. It also includes the neutrinos, which are the massless particles that appear among the decay products of the neutron, as well as in other interactions.

Particles that have integer spin—0, 1, 2, and so on—are called bosons, after Indian physicist S. N. Bose. The only boson we have encountered so far is the photon, the particle associated with electromagnetic radiation.

Fermions and bosons are truly distinct classes of particles in the sense that no interaction has ever been seen that can convert one into the other. This distinction, as we shall see, persists back to the earliest stages of the Big Bang.

CLASSIFICATION BY DECAY PRODUCT

When an elementary particle disintegrates, the decay products can include particles that are nominally stable, such as the electron, proton, photon, or neutrino, or they can include particles that will themselves eventually decay. It is clear, however, that if we watch long enough and let all of the decay products and the products of the decay products disintegrate, we will eventually wind up with a collection of relatively stable particles. Now there are two possibilities. Either this collection will include an excess of protons over antiprotons (or vice versa) or it will not. If it does not, the original particle is called a meson (literally, "the intermediate one"); otherwise, it is called a baryon (literally, "the heavy one").

The term *meson* was coined in the 1930s because the only particles known at that time that fit our definition were intermediate in mass between the electron and the proton. Let us look at two important examples of mesons. The first one discovered was denoted by the Greek letter *mu* (μ) and was called the mu-meson or muon. It has a mass of .105 GeV, about two hundred times the mass of the electron and one-ninth that of the proton. It has spin ½ and decays in about a microsecond into an electron, a neutrino, and an antineutrino. All of these decay products are stable, and the collection contains no protons, so it is clear that the particle is a meson.

Similarly, a type of meson denoted by the Greek letter *pi* (π) and called the pi-meson or pion was discovered in the late 1940s. The three members of the pi-meson family have positive, negative, and neutral electrical charges, respectively, and masses around .14 GeV. The two charged members of the family decay in 10^{-8} second into a muon and a neutrino, while the neutral member decays in 10^{-16} second into a pair of photons. The muons created by the decay of the charged pions decay themselves, of course, so the collection of decay products for a negatively charged pion, for example, would consist of an electron and a collection of three neutrinos. This collection contains no proton, so the pion definitely meets our definition of the meson.

In both of these examples, the mass of the meson is less than that of the proton. For such particles, there simply is not enough energy in the mass of the meson to produce a proton among the decay products—they are mesons by default. There are, however, many known mesons whose mass exceeds 1 GeV, and for such particles a proton could, in principle, be seen among the decay products. The fact that one is not seen, therefore, tells us something fundamental about the nature of the particle.

The term baryon is applied to a particle whose decay eventually results in a proton plus other stable particles. Obviously, every baryon must be at least as massive as the proton. While there are about as many baryons in particle lists as there are mesons, we will describe only one here. This is a family of particles designated by the Greek letter delta (Δ). This family has four members, whose electrical charges range from two positive units down to one negative unit and which have a mass near 1.2 GeV. The delta decays in about 10^{-23} second into a proton or a neutron and a pion. Thus, its final decay products will be a proton and collection of electrons, positrons, and neutrinos. The presence of the proton, of course, is what puts this particle into the baryon category.

CLASSIFICATION BY INTERACTION

Most of the elementary particles, even those with very short lifetimes, can exist inside of the nucleus of atoms. The delta, for example, has a lifetime of 10^{-23} second, but if it is traveling near the speed of light (3×10^{10} cm/sec), it can travel 10^{-13} cm before it decays. This is typical of the distances between particles in a nucleus, so it is possible for the delta to serve as a means of communication between different parts of the nucleus during its short lifetime. We say that the delta "participates" in the interactions that hold the nucleus together. All particles that share this characteristic are called hadrons ("strongly interacting ones"). This term comes from the fact that physicists call the force that holds the nucleus together the strong interaction.

Almost all of the known particles are hadrons, and the study of hadronic matter has been a major concern in physics since the early 1960s. A few particles, however, do not fall into this category. The photon, for example, is not a hadron. It is, to all intents and purposes, in a class by itself (a point that will become very important later).

The other nonhadronic particles are those that are involved in the process of radioactive decay. They are called leptons ("weakly interacting ones"). The most familiar lepton is the electron. It does not exist in the nucleus of an atom, but instead circles the nucleus in orbit. It is one of the end products of the relatively slow decay of particles like the neutron and muon. It turns out that there are two other particles like the electron. We have already encountered one—the mu-meson. It does not exist in the nuclear maelstrom either, but the key point is that except for its greater mass, the muon is just like the electron. It has spin ½ and participates in the radioactive decay process. In fact, for a long time this puzzled physicists. Why had nature made the electron and then done the same job over again two hundred times heavier?

A good deal of the urgency was removed from this question in 1975 by researchers at Stanford University. A third lepton, denoted by the Greek letter tau (τ), was found. Its mass (1.8 GeV) was thirty-six hundred times that of the electron and seventeen times that of the muon, but that was the only difference between the tau and these particles. It seems as if nature, for some reason, has made many massive leptons.

The other particles included in the lepton category are the neutrinos. We know that a neutrino is emitted in conjunction with mu-mesons when the pion decays and also in the decay of the neutron. It turns out, however, that the neutrinos emitted in these two processes are not identical. If we take the neutrino from pion decay and allow it to strike a target, it will produce mu-mesons but not electrons. Conversely, electrons from processes like neutron decay cannot produce mu-mesons. Presumably, neutrinos emitted in conjunction with the tau would behave in the same way. There seems to be, in other words, a grouping of the electrons and neutrinos into pairs: each massive particle is associated with one, and only one, type of neutrino. These groupings go by the name *generation*. All of the known leptons are shown in the table on the next page.

This concludes our discussion of the major categories into which elementary particles are grouped. You will note that al-

Generation	*Massive Particle*	*Neutrino*
First	Electron	e-neutrino
Second	Mu-meson	μ-neutrino
Third	Tau-meson	τ-neutrino

though we have talked about such properties as spin and decay, we have not paid a great deal of attention to electrical charge. The reason for this is quite simple: at the kind of temperatures that existed around the beginning of the particle era, the interactions between particles were so energetic that all members of a family can be thought of as having the same behavior, simply because the differences in behavior due to electrical effects are so small as to be negligible. Thus, for example, we will speak of the pion without bothering to specify which of the three pi-mesons (positive, neutral, or negative) we mean. The only place that electrical charge is important is in dealing with conservation laws, a subject to which we now turn our attention.

The Conservation Laws of Particle Physics

Interactions between elementary particles do not occur at random but are governed by a set of general principles known as conservation laws. A physicist says that a quantity is "conserved" if it does not change during a particular physical process. We have already alluded to one such principle, the conservation of energy. The energy of an isolated system cannot change, and we have used this fact to show that a particle cannot produce decay products heavier than itself. If it did, then the mass energy of the decay products would be greater than the mass energy of the original particle, which would mean that the total energy of the system would have increased during the decay. This would be a direct violation of energy conservation and, therefore, could not occur in nature. This example illustrates one general use of conservation principles. They tell us what reactions can proceed in nature and allow us to explain why some reactions (those that violate the laws) are never seen.

Another extremely important conservation law governs electrical charge. If you add up the electrical charges of the particles that enter a reaction and compare it to the sum of the charges of the particles produced, the two numbers must tally. For example, the decay of the neutron can be written symbolically as

$$n \rightarrow p + e^- + \nu$$

where the Greek letter nu (ν) represents the neutrino. The total charge in the beginning is zero, since the neutron is electrically neutral. The total charge at the end is +1 (proton) −1 (electron) +0 (neutrino) = 0, exactly as it should. Although the conservation of energy would allow a neutron decay like

$$n \rightarrow p + \nu + \nu$$

the conservation of charge rules it out, since it would require the total charge of the system to change from 0 to +1 during the decay. One way to think of this result is to note that when we say that charge conservation rules out the neutron decay shown above, we are saying that the time required for this decay to proceed is infinite.

Charge is an example of something physicists call an internal symmetry of a system. The electrical charge of a particle has nothing to do with how fast it is moving or whether it is spinning on its axis. It is a totally separate quantity that obeys its own set of rules. As far as we know, charge conservation is an exact and universal law, and there are no known violations anywhere in nature. There are, however, other internal symmetries that are not so universal. Instead of providing a flat prohibition against a given reaction, they simply require that the reaction proceed much more slowly than one might have expected.

The first of these symmetries to be encountered showed up in some reactions initiated by cosmic rays in 1947. A particle created in the collision of a proton with a nucleus was observed to live for a very long time before decaying—around 10^{-10} second. Denoted by the Greek letter lambda (Λ) this particle was heavier than the proton, and physicists were puzzled as to why it did not decay much more quickly than it did. On first glance, one might expect such a particle to decay in something characteristic of the nuclear communication times—10^{-23} second or so—yet here was a particle with a lifetime a trillion times longer. As more species of these particles were seen, they were given the name strange particles, and eventually physicists began to think of "strangeness" in much the same way they thought about electrical charge. Strangeness (or the lack of it) became an intrinsic attribute of particles, just like electrical charge, except that strangeness was not strictly conserved in particle interactions. Unlike charge conservation, the law of strangeness conservation seemed to decree that when strangeness conservation was violated (that is, when the strangeness was different after an interaction from what it was before), the interaction time was greatly increased, but not increased all the way to infinity. Thus, in the decay of the lambda,

$$\Lambda \rightarrow p + \pi^-$$

electrical charge is conserved, but strange charge is not. The strangeness of the particles on the right is 0, while the strangeness of the lambda, by convention, is -1. Thus, the law of strangeness conservation tells us that this reaction should proceed slowly, as indeed it does.

In 1974 two independent teams of physicists discovered a new particle that displayed yet another kind of internal symmetry. This quantity was called charm (particle physicists have gotten into the disturbing habit of finding "cute" names for quantities they introduce). Charm can also be thought of as being analogous to electrical charge, and just as entire families of strange particles were discovered, so too have many charmed particles been seen in the laboratory.

Two more attributes like charm and strangeness are supposed to exist. They are denoted by the letters *b* and *t,* and you can choose as names either *bottom* and *top* or *beauty* and *truth.* Particles carrying the *b* charge have been discovered, although not in great numbers. The *t* particles have not yet been seen; they are supposed to be about ten times more massive than the proton and therefore a bit above the capacity of present accelerators to produce.

Thus, we can define a particle according to the various "charges" it carries. For example, the proton has electrical charge $+1$, strangeness, charm, *b*, and *t* equal to 0. This string of numbers, together with the number designating the spin, are called the quantum numbers of the particle being described. The quantum numbers of antiparticles are simply the negative of those associated with the particle.

Quarks

If we return to the thought with which we began this discussion of elementary particles—that the fundamental building blocks of matter ought to be simple and few in number—it is clear that the kind of proliferation of elementary particles that took place between 1950 and 1970 just will not do. Always before when a complex situation like this was found in nature, it was discovered that if one descended to a deeper level of reality, the complexity was replaced by a new simplicity. The idea of atoms, as well as the proton-neutron-electron picture of matter, represent this sort of process. It should come as no surprise, then, to learn that as early as 1964 physicists were proposing that the multitude of elementary particles were actually composed of still more basic particles called *quarks.* The name itself is an allusion to a line in James Joyce's *Finnegans Wake*—"Three quarks for Muster Mark." The allusion made sense when the theory was introduced because it was thought originally that there were three different kinds of quarks. The name has stuck even though the number of quarks has grown since then.

The idea is that protons, neutrons, pions, and all other hadrons are simply different arrangements of the basic quark building blocks. The baryons are made of three quarks, while the mesons are made of a quark and an antiquark. The most striking feature of the quarks is that they have a fractional electrical charge. The quantum numbers of the original three quarks are given in the following table.

Kind of Quark	*Spin*	*Electrical Charge*	*Strangeness*
u	$1/2$	$2/3$	0
d	$1/2$	$-1/3$	0
s	$1/2$	$-1/3$	-1

The three antiquarks, of course, have the opposite electrical and strange charges from those in the table. It is customary to denote the antiquarks by writing a bar over the letter. For example, the anti-u quark is written $\bar{u}$.

Putting together these quarks to mark elementary particles is a simple exercise, somewhat akin to building a structure from a child's blocks. The proton, for example, must have the structure shown in Figure 6. It is composed of two *u* quarks and one *d* quark, two of which are spinning in the opposite direction from the third. The total spin of this object (the sum of the spins of the quarks) would be $1/2 + 1/2 - 1/2 = 1/2$, while the total charge would be $2/3 + 2/3 - 1/3 = 1$. These, of course, are the quantum numbers of the proton.

A pion, on the other hand, would have a quark structure shown in Figure 7—a *u* quark and an anti-*d* quark spinning in opposite directions. The spin ($1/2 - 1/2 = 0$) and electrical charge ($2/3 + 1/3 = 1$) of this combination are precisely the values needed for the positively charged member of the pi-meson family.

In point of fact, all of the particles with zero strange charge can be made by combining the u, $\bar{u}$, d, and $\bar{d}$ quarks. To produce a particle with strangeness, however, it is necessary to include the *s* quark in the structure. For example, if we replaced one of the *u* quarks in the proton with an *s* quark, we would have a particle with spin $1/2$, electrical charge $2/3 - 1/3 - 1/3 = 0$, and strange charge $0 + 0 - 1 = -1$. These are precisely the quantum numbers of the lambda particle. The decay of the lambda into a proton and pion involves two processes at the quark level: the *s* quark must be converted into a *u* or *d*, and a quark-antiquark pair must be created. This process is shown in Figure 8. Thus, the slow decays associated with strangeness change are understood at the quark level as being caused by the requirement that one of the quarks in the original particle has to be converted into a different type of quark.

This understanding of strangeness indicates why it was that when a new long-lived particle was discovered simultaneously at Brookhaven Laboratories in New York and the Stanford Linear Accelerator Center in California, the slow decay was taken as

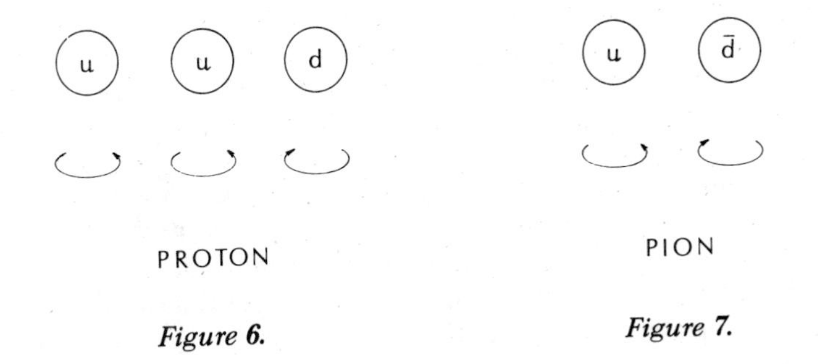

Figure 6. *Figure 7.*

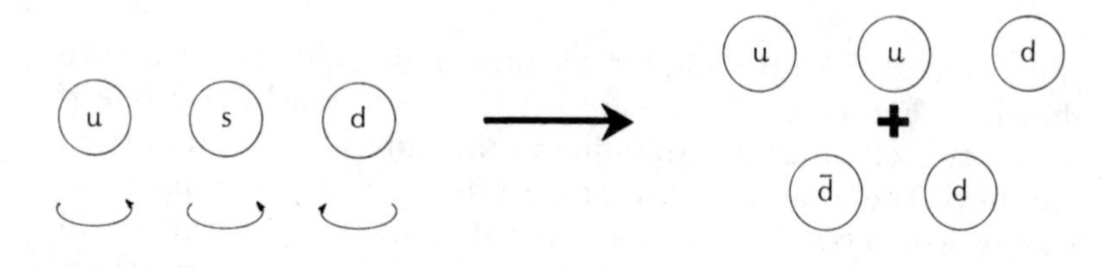

Figure 8.

evidence for a fourth quark bearing a type of change that came to be termed charm. The quarks that have been added to the list since then are given in the following table.

Quark	Charge	Strangeness	Charm	*b*	*t*
c	$2/3$	0	1	0	0
b	$-1/3$	0	0	1	0
t	$2/3$	0	0	0	1

All of these quarks have spin $1/2$, and for each of them there is an entire family of particles formed by replacing *u* and *d* quarks in ordinary particles with *c, b,* or *t* quarks. The structure of the lambda, formed by replacing a *u* quark in the proton with an *s* quark, is an example of this process.

It is interesting to note that, just as there are six leptons arranged in three generations, there are also six quarks. The quarks can also be arranged in generations as follows:

Generation	Quarks
First	*u, d*
Second	*s, c*
Third	*b, t*

The analogy between quarks and leptons will be important later.

Finally, there are two other terms that are used when speaking of quarks. We have talked about the kind of quark when we wish to distinguish, for example, between *s* and *u* quarks. Following the "cute" tradition, physicists speak of the flavor of the quark when making this distinction. Thus, the *u* and *d* quarks are in the same generation, but are of different flavors.

For technical reasons we need not go into deeply here, physicists feel that there must be yet another internal symmetry associated with the quarks, a kind of attribute that goes by the name *color.* Unlike charm and strangeness, this particular attribute has no immediate experimental consequence that can be easily singled out. For the moment, therefore, we will simply note that in the grand unified theories, each of the kinds of quarks included in the tables (*u, d, s, c, b,* and *t*) is actually a family of three quarks, each one carrying a different color "charge." This is similar to the pion, where the π is actually a family of three particles distinguished by having different electrical charge. You can think of the quarks as being red, white, and blue (my own sentimental favorite) or red, green, and blue (the primary colors), since there is no generally accepted chromatic terminology at the present time. The main thing to keep in mind is that there is no suggestion in this terminology that the quarks are covered with some sort of paint. Color, as applied to quarks, simply refers to an intrinsic property of the particles, just as electrical charge and strangeness do.

Summary

The properties of the elementary particles that can be used to classify them are mass, spin, and decay. The masses of elementary particles are measured in units called GeV (gigaelectron volts), with 1 GeV being slightly more than the mass of the proton. Particles can be thought of as small spheres rotating on their axes. The laws of quantum mechanics require that the rate of spin for a given particle must have specified values. It is customary to describe the rotation of a particle by giving the number by which a basic spin rate must be multiplied to give the actual spin of the particle. Thus, a particle rotating at ½ its basic spin rate is said to have spin ½, and so on.

Almost every elementary particle is unstable in the sense that if it is left alone long enough, it will disintegrate, producing a new set of particles called decay products. The lifetime of a particle is the average time it can live before it decays; it can be as short as 10^{-23} second (the delta) or as long as billions of years (the uranium nucleus).

Based on these and other properties, particles can be classified in many ways. Some categories that will be important are the following:

fermions	*particles of spin ½, 3/2, 5/2 . . .*
bosons	*particles of spin 0, 1, 2, 3 . . .*
baryons	*particles whose eventual decay products include a proton*
mesons	*particles whose eventual decay products do not include a proton*
hadrons	*particles that participate in the interaction that holds the nucleus together*
leptons	*particles that do not participate in the interaction that holds the nucleus together*

Of these classification categories, the leptons are simplest to describe, since there are only six of them. They are all spin ½ fermions, the most familiar being the electron. Associated with each massive lepton like the electron is a massless particle called the neutrino. We speak of grouping the leptons in generations. Thus, the first generation of leptons consists of the electron and its associated neutrino, the second generation is the mu-meson with its neutrino, and the third is the tau-meson with yet another neutrino. The mu- and tau-mesons are similar in every respect to the electron except for the fact that they are heavier.

The behavior of elementary particles is governed by a set of conservation laws. These are laws that state that in any interaction certain quantities stay the same (are conserved). Electrical charge is one example, since we know of no cases in which the total charge of an isolated system can change. This law is associated with the existence of electrical charge as a property of elementary particles. There are other chargelike attributes carried by elementary particles besides the familiar electrical one. These other attributes (or internal symmetries, to use the technical term) are called strangeness, charm, beauty, and truth. Unlike electrical charge, these other "charges" are not strictly conserved, but reactions in which they change take place much more slowly than reactions in which they do not.

The hundreds of known elementary particles are thought to be made up of still more elementary building blocks called quarks. Baryons are composed of three quarks, and mesons, of a quark and an antiquark. There are six kinds of quarks, all of which carry a fractional electrical charge. In addition, the s, c, b, *and* t *quarks each carry one unit of strange, charm, beauty, or truth charges, respectively. Finally, it is now believed that there is yet another internal symmetry called color and that each kind of quark actually represents a family of three particles that are identical in all respects except that they are of different colors. Thus, it takes eighteen quarks (six kinds of three colors each) and eighteen antiquarks to make the world.*

Chapter 5

The Fundamental Forces

Every body continues in a state of rest, or of uniform motion in a right line, unless it is compelled to change that state by forces impressed upon it.

ISAAC NEWTON
Principia Mathematica

Gravity

We all have constant contact with one of the fundamental forces of nature—the force of gravity. Yet as familiar as this force is in our everyday existence, it was not until Isaac Newton formulated his law of universal gravitation in 1666 that a very important fact about gravity was understood. Up until the time of Newton, it was generally believed that there was not one but two forces of gravity operating in the universe. One of these, the one seen on the earth, is the familiar force that causes bodies to fall when they are unsupported; the other force, which did not go by the name gravity, operated in the heavens, impelling the planets and stars around on their crystal spheres. In pre-Newtonian thought, the two phenomena involved were so patently different that it was obvious that the two forces involved were totally distinct and bore no relation to each other.

Newton's great achievement was to show that there was only one gravitational force in the universe and that the motion of the planets and the fall of an apple are both simply manifestations of a single universal effect. In modern language, we say that Newton "unified" the two forces of gravity and that the Newtonian theory of gravitation is a "unified theory." Since we often speak of gravitational fields when talking about this subject, the term *unified field theory* could also be applied.

The way that Newton arrived at his law of universal gravitation is interesting. He started by thinking about a simple projectile thrown on the surface of the earth. The object would travel a certain distance (a few yards, for example) before the force of earthly gravity pulled it back to the ground. If the projectile were thrown more forcefully, it would travel farther before striking the ground; also, the faster it was moving, the farther it would go. Taking this process to an extreme, Newton imagined throwing something so fast that it would be able to go all the way around the earth and return to the spot from which it was thrown. With just the right velocity, it would come back to this point moving at precisely the same speed with which it had started, in which case it would go around again. The similarity between the projectile circling the earth and the moon in its orbit was all that Newton needed to realize that there was no reason to have two separate gravitational forces operating in the universe. The same force, the one we now refer to as gravity, could explain both the motion of objects near the surface of the earth and of bodies in orbit in the solar system. From that time forward, the artificial distinction between heavenly and earthly phenomena was dropped, and it was assumed that the same laws operated everywhere.

For our discussion, the two important facts about gravity have to do with its range and its strength. The range of a force is the distance over which that force can be applied. For gravity, the range is infinite. The most distant galaxies are exerting a force on you as you read this, although the force is so small that it could never be measured. According to Newton, every body in the universe exerts a force on every other body, no matter how great their separation.

The strength of a force is something that is usually discussed in relation to other forces, rather than in absolute terms. Although it may be hard to believe after you have helped a friend move a sofa up to a third-floor apartment, gravity is by far the weakest of the fundamental forces. The reason it dominates our lives the way it does is that we spend our days on the surface of a huge mass (the earth) that functions as a gigantic generator of gravitational force. However, the fact that you can pick up a nail with a magnet shows that even the entire earth pulling on one side cannot counteract the magnetic force exerted by something that can be held in your hand.

Electromagnetism

The Greeks were well aware of the fact that if certain common objects, such as a piece of amber, were rubbed, bits of material would cling to them. Today, we would say that the Greeks knew about static electricity. (Indeed, the word *electron* means amber in Greek.) The Greeks were also acquainted with the force of magnetism. Neither of these forces, however, was studied systematically until the end of the eighteenth century, although magnetism had been used extensively in the design of navigating compasses. We know that there are two kinds of electrical charge, which we call positive and negative, and that like charges attract, while unlike charges repel, each other. We also know that if two magnets are brought together, their north poles will repel each other, but the north and south poles will attract each other. As was the case with earthly and heavenly gravity, there seems to be little connection between the magnet that holds notes to your refrigerator door and the static electricity that causes a spark to jump from your hand after you have walked across a rug on a dry day. One of the great achievements of nineteenth-century physics was to show that these two forces, seemingly so different from each other, are in fact nothing more than different aspects of a single force—electromagnetism.

The first step to this realization was made by the Danish physicist Hans Christian Oersted in 1819. Quite by accident, he noticed that when he connected a wire across the poles of a battery so that an electric current ran, the needle of a nearby compass was deflected. Subsequent investigation showed that whenever an electric charge moved it was capable of producing a magnetic force. This was true whether the charge involved was a single particle moving freely in space or the large number of particles that constitute an electric current in a wire. It was not long before the converse relation was established. It was shown that a moving magnet was capable of exerting an electrical force (that is, of making an electrical charge move). Thus, a connection between the two seemingly disparate phenomena was established. This connection represents a slightly more abstract type of unification than that established for gravity by Newton. Electrical and magnetic phenomena are still perceived as being different, but these experiments show that the differences are not really fundamental. Therefore, physicists were led to think of electricity and magnetism as two distinct manifestations of a single underlying process.

The final touch was put on this process of unification by the Scottish physicist James Clerk Maxwell, who published what is now taken to be the definitive theory of electromagnetism in the form of four equations (known, appropriately enough, as Maxwell's equations). In these equations electricity and magnetism enter as equal and inextricably intertwined partners. Indeed, the equations showed that in free space, far from any material, it is possible to interchange electrical and magnetic fields without producing any change whatever in the basic equations.

As was the case with gravity, the range of the electromagnetic force is infinite. A charged particle on one side of the galaxy is, in principle, capable of exerting a force on a charged particle on the other side of the galaxy, although the force would be extremely small because of the distance between them. While electromagnetism is a stronger force than gravity, as we have argued, in the hierarchy of strengths of the fundamental forces its strength is intermediate—neither the strongest nor the weakest.

The most important role of electromagnetism in nature is to serve as the force that binds orbiting electrons to nuclei in atoms. It is also the electromagnetic force that binds separate atoms together into molecules and holds them together in solids. Thus, in a very real sense, it is the electromagnetic force that is responsible for most of the things we see in the world around us.

Strong and Weak: The Subatomic Forces

The development of nuclear physics in the late nineteenth and early twentieth centuries led to the discovery of the other two fundamental forces—the strong and weak forces (or interactions). These forces will be less familiar, since they are not manifested directly in familiar processes. Nevertheless, they are real and can be studied.

The existence of the strong force is actually implicit in the picture of the atom, in which all of the positive charge is concentrated in the tiny volume of the nucleus. We know that like charges repel each other, so if two protons in a nucleus are really only 10^{-13} cm apart, they must be pushing away from each other with a tremendous force. If, for example, we scaled the protons up to the size of basketballs, the scaled-up repulsive force between them would be so large that they would fly away from each other even if the basketballs were encased in a cube of solid steel. The natural question, then, is how the nucleus can stay together in the face of the disruptive electrical repulsion. Clearly, there must be some other force operating here, some force capable of overcoming the repulsion and binding the nucleus together. This is the strong force, the study of which has been a major concern of twentieth-century science.

As the name implies, the strong force is the most powerful of the known interactions. Its range, however, is quite limited. In fact, two particles are capable of exerting this force on each other only if they are inside the same nucleus. If they are more than roughly 10^{-13} cm apart, the strong force will cease to operate. Thus, protons in the nuclei of two different atoms will not be capable of interacting with each other via the strong force.

The weak force is best known for its role in radioactive decay. Natural radioactivity was discovered and elucidated by Antoine Henri Becquerel, Pierre and Marie Curie, and others at the end of the nineteenth century. In the 1930s scientists learned how to create new species of unstable nuclei, and man-made radioactivity has, for better or worse, become part of our lives. If you consider the nucleus of a uranium atom, none of the other three forces could explain how it could emit two protons and two neutrons, transforming itself into a nucleus of thorium in the process. This sort of slow decay must be governed by an as yet undefined force, the weak force. Like the strong force, the weak interaction is not something with which we have extensive everyday experience. It has a very short range—only about 10^{-15} cm (about 1 percent of the distance across a proton)—and is intermediate between electromagnetism and gravity on the scale of strengths.

The Four Forces: An Overview

It would appear that the situation at the present time is that we can explain everything we see in nature in terms of four fundamental forces: gravity, electromagnetic, strong, and weak. The first two forces actually represent unifications of other forces which had previously been thought to be distinct and fundamental in their own right. The following table is a summary of the properties of the forces.

Properties of the Fundamental Forces

Force	Strength (relative to strong)	Range (cm)
Strong	1	10^{-13}
Electromagnetic	1/137	infinite
Weak	10^{-5}	10^{-15}
Gravity	6×10^{-39}	infinite

The first thing that strikes us when we look at the table is the enormous disparity in the strengths of the forces. They range over thirty-nine orders of magnitude, a number so large as to be almost incomprehensible. The ranges of the forces go from something smaller than a proton to something large enough to encompass the entire universe. In the face of these facts, it would seem reasonable to suppose that the successful unification we realized earlier must surely stop at this point. How could forces so different actually have an underlying identity?

But are the differences between the forces as great as they seem? Are they, for example, greater than the once imagined difference between the force that makes an object fall on the earth and the force that holds the planets in their orbits? I would argue that the unifications achieved before the twentieth century were at least as difficult as the task of unifying the forces in the table. This is particularly true of the Newtonian theory of gravitation, a case where there was virtually no theoretical precedent for synthesis.

Besides, there is something inherently inelegant in having to have four different forces (and consequently four different theories) to explain four different types of natural phenomena. It is clear that there has to be at least one force in nature, otherwise there could be no interactions at all. But what is special about the number 4? If we accept the four forces as being truly fundamental and independent, we are immediately faced with the problem of explaining why there are four forces and not three or five or any other number. This apparent lack of elegance is very troublesome to physicists, who like to believe that nature must be fundamentally simple and beautiful.

Of course, the fact that there have been successful unifications in the past does not mean that the unification of the four forces will be easy. Men of the stature of Albert Einstein spent much of their lives trying to develop the ultimate unified theory and failed. Einstein started with gravity and tried to unite the other forces with it. As it turns out, this is the hardest way to go at the problem, a fact that in retrospect helps to explain Einstein's failure. We shall see that we have to start with the forces which act between elementary particles—the weak, electromagnetic, and strong forces—if we are to arrive at a unified theory. This, in turn, means that we must have some understanding of the way that modern physicists view these forces—a subject to which we now turn our attention.

Force as Exchange

When Newton first gave his definition of force, his main concern was to show how to recognize when a force was acting. The commonsense notions of pushing and pulling were what people had in mind when they thought about forces. This led to some interesting, albeit pointless, debates on the question of "action at a distance." How could the sun affect the motion of the earth if the two are not touching each other?

The answer to this old question and to many others can be found in a new way of thinking about forces that arises from the study of elementary particles. All four of the fundamental forces are now considered to arise from the exchange of elementary particles. A simple analogy may make this concept easier to understand. Suppose two ice skaters are moving toward each other along parallel tracks, as shown in Figure 9. At some point one skater throws a snowball at the other, recoiling slightly as he does so. Some time later the snowball strikes the second skater, causing him to recoil in the opposite direction. The net effect, then, is that the two skaters' paths are deflected when they come near each other. Applying Newton's definition, we would say that a force had acted on the skaters, since they were no longer moving in the same direction as they had initially. In fact, since the skaters were deflected away from each other, we would say that a repulsive force had acted. It is somewhat harder to find a simple

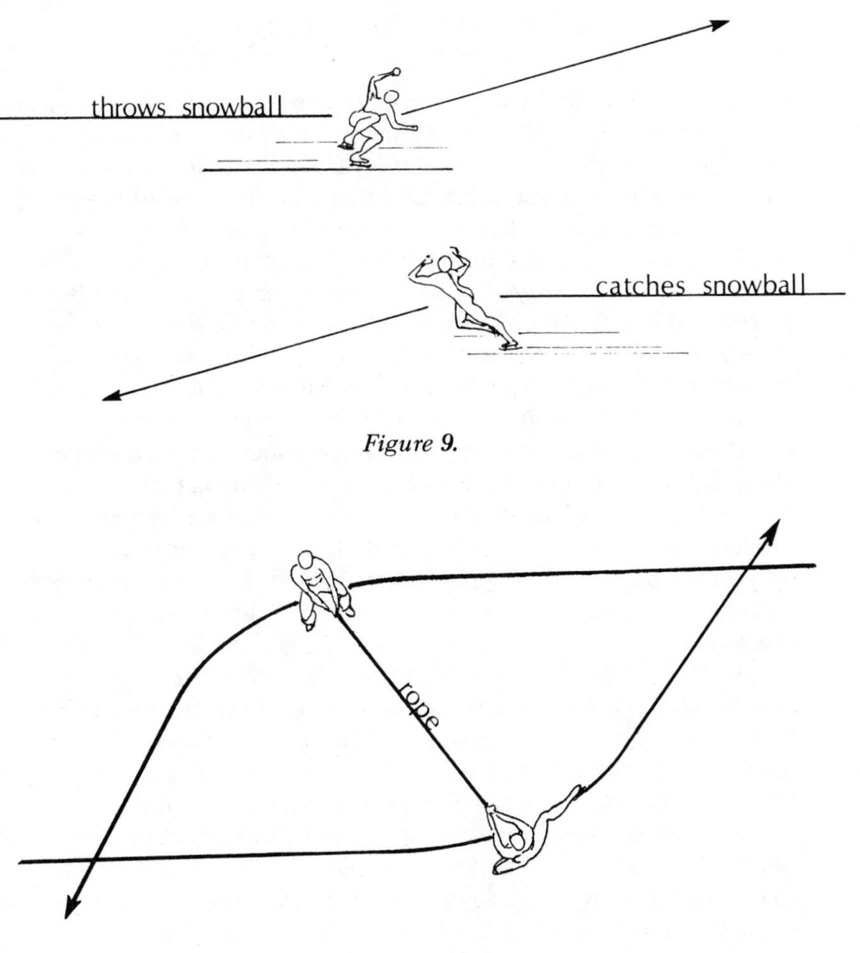

Figure 9.

Figure 10.

picture that associates an attractive force with an exchange, but you can get some notion of how such a situation could come about by imagining the first skater throwing one end of a very lightweight rope to the other. If both held tightly to their ropes, their paths would be as shown in Figure 10. In this case we would speak of an attractive force acting between the skaters. We shall see that at the level of elementary particles, exchanges generate both attractive and repulsive forces.

That such a situation might occur in the strong interaction arises because of an important feature of the subatomic world—Heisenberg's uncertainty principle. The version of this principle that we wish to consider here concerns the law of conservation of energy. Since we know that energy is required to produce a massive particle, we might expect that a strict interpretation of this law would require that a proton sitting by itself in a vacuum could not emit a particle like the pion. To do so would require that the mass energy of the pion appear out of nothing. This interpretation, however, ignores one very important fact about the conservation of energy (or any of the other great principles of physics): the validity of any natural law is established by experiment, and a law is only as good as the experiments used to confirm it. We can imagine that, by some process, the proton could create a pion but would reabsorb it in a time so short that we could not detect the change in energy associated with the creation. If this happened, we could not say that the law of conservation of energy had been violated, because we could detect no evidence of violation. It would be as if someone came into your office and borrowed a book, but returned it before you were aware of its absence. You could not say the book had been missing.

In the traditional view of physics, we do not consider processes like the creation of the pion from a proton because we say that we could, in principle, observe the proton on as small a time scale as we wished, so that we could always make the observation time smaller than the time required to reabsorb the pion. In quantum mechanics, however, this sort of constant surveillance is not possible. The uncertainty principle takes account of the fact that every measurement made on a particle affects that particle, so that there is a limit on how accurately we can know two separate things, such as the particle's energy and the time at which it possesses that energy. In essence, the uncertainty arises because the more accurately we measure one quantity, the more we disturb the system and the less information we can obtain with our second measurement.

If we call ΔE the uncertainty in our knowledge of the energy of a system and Δt the uncertainty in our knowledge of the time at which the system has that energy, then the mathematical statement of the uncertainty principle is

$$\Delta E \times \Delta t > h$$

where $>$ is the mathematical symbol for "greater than" and h is a number known as Planck's constant (after the German physicist Max Planck, one of the founders of modern physics). What this statement tells us is that if we measure the energy of a system very precisely, so that ΔE is very small, then Δt must be very large. You can think of this fact in the following way: To determine E precisely, we have to observe the system for a long time; the result of such a long-term measurement will be an accurate determination of the average energy of the system over that period, but it will be impossible to tell exactly when the system actually had that value of the energy, which means that Δt must be very large.

With this knowledge in mind, let us go back and think about the proton and the pion. We have said that the pion cannot appear out of nothing—a statement which is undoubtedly correct. Let us, however, put the question in a slightly different way. For how long a time could the uncertainty in our knowledge of the energy of the proton be so large that we could not, according to the uncertainty principle, tell if a pion had been created? In other words, what if the proton suddenly created a pion out of nothing but reabsorbed it too quickly for us to detect the pion's presence? Such a process would not violate the conservation of energy, since there would be no experiment that could be done, even in principle, that could show the energy of the proton to change spontaneously.

In the system of units where length is measured in centimeters and mass in grams, h has the value 6.6×10^{-27}, the pion mass is about 1.8×10^{-25} gm, and c is 3×10^{10} cm/sec. According to the uncertainty principle, the uncertainty of the energy of the proton could be as large as the pion mass energy ($m_\pi c^2$) for a time Δt if we had

$$(m_\pi c^2)\Delta t > h$$

or

$$\Delta t > 4 \times 10^{-23} \text{ sec}$$

In other words, in the quantum world the proton can create a pion out of nothing, provided that the pion does not stay around for more than this amount of time. Like Cinderella, if the pion gets home in time, no one will be able to tell that anything is amiss. A particle that exists for only the fleeting instant allowed by the uncertainty principle is called a virtual particle.

The connection between a virtual pion and the strong force

can be seen if we ask how far a virtual pion could travel in the time we have calculated. Since the pion can travel, at best, at the speed of light, the distance it could travel would be $c \times \Delta t$, or about 10^{-12} cm. This is roughly the diameter of a medium-sized nucleus. In other words, two protons in a nucleus could exchange a pion between them in much the same way that the skaters exchanged snowballs or ropes in our analogy. Provided that the first proton emitted the virtual meson and the second absorbed it in a time less than 4×10^{-23} sec, such a process would not violate any known law of physics.

Just as the exchange of snowballs between skaters could be thought of as causing a force to act, so too can the exchange of virtual pions be thought of as generating a force between particles in the nucleus. And just as a snowball can only be thrown so far, so that there is a limit to the range of the "snowball force," a virtual pion can travel only a short distance without violating the principle of energy conservation. In both cases, the force generated by the exchange has a limited range. In fact, the only difference between the snowball and the pion is that the former produces a repulsive force, while the latter produces an attractive one.

This pion-exchange force, which tends to bind particles together into a nucleus, is what we identify with the strong interaction. In a sense, the virtual pions form the glue that holds the nucleus together despite the powerful forces of repulsion between the protons. One way of picturing a nucleus, then, is to imagine the protons and neutrons sitting in a nucleus emitting and absorbing virtual pions. Diagrammatically, we show this process as in Figure 11, with a simple two-particle nucleus on the left and a more complex nucleus on the right.

Now the argument we made for the existence of virtual pions need not be restricted to any one particular species of particle. Any particle can exist in the virtual state, but the more massive a particle is, the greater the ΔE, the energy uncertainty,

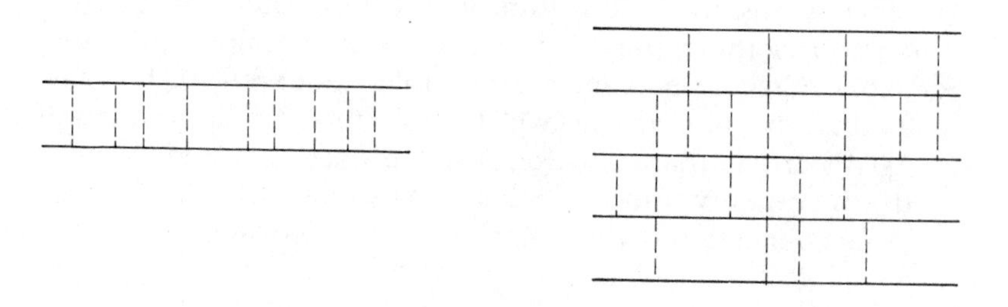

Figure 11.

must be. This implies that Δt, at the time during which the particle can exist, must be shorter than the time allotted to the pion and hence that the distance the particle can travel (a number we identify with the range of the force) must be smaller. The heavier the exchanged particle, the shorter the range of the force. The pion, being the lightest hadron, will be associated with the long-range part of the strong force, but other, heavier particles will come into the picture when the protons are closer together. In fact, it is customary these days for nuclear theorists to include many particles heavier than the pion in their calculations.

Thus, we see that one important property of the forces—their range—can be understood in terms of the exchange mechanism. What about the other important property—the strength? If we think back to the analogy of the skaters and the snowballs, we realize that the amount of deflection of the skaters' paths—a quantity we would associate with the strength of the force—depends on what is exchanged. If the skaters threw water at each other or blew a stream of air through a pipe, the amount of deflection would be much less than that experienced for a snowball. Clearly, the strength of the force we associate with the exchange depends on the way what is exchanged interacts with the targets.

In our particle-exchange picture, this added dimension of the process is usually taken into account by assigning a number to express the strength of the interaction at the points where the particle is emitted and absorbed. These points correspond to the vertices labeled with a small g in Figure 12.

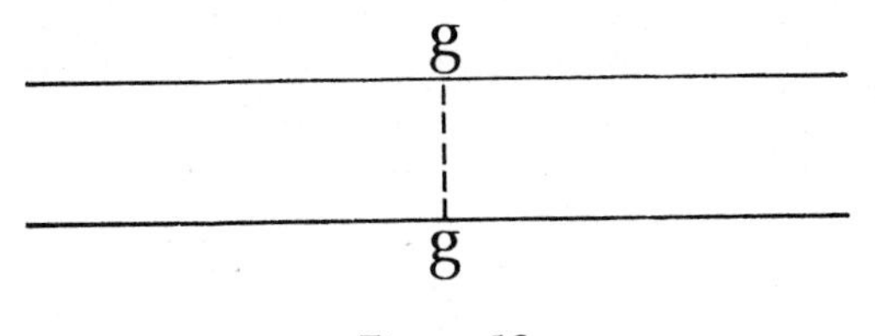

Figure 12.

The number, which we will denote by g, is called a coupling constant. It is this number that we have referred to loosely as the strength of the forces up to this point. The coupling constants associated with the fundamental forces will be very important in the development of the unified theories we will describe in the next few chapters.

The Fundamental Forces as Exchanges

Two of the fundamental forces, electromagnetism and gravity, have an infinite range. From our knowledge of the exchange picture, we know that the lighter the mass of the exchanged particle, the longer the range of the association with the force will be. It is not hard to extrapolate this idea of increasing range to the limit of a particle whose mass is zero. In this case, going through the kind of argument we just followed for the exchange of a meson shows that the range of the force would be infinite. Thus we conclude that the particles being exchanged in electrical and gravitational interactions must have zero mass.

You will recall that when we talked about classifying particles in Chapter 4, we noted that the photon was in a category by itself. It turns out that the photon is precisely the zero-mass particle whose exchange generates the electromagnetic force. When you lift a nail with a magnet, huge numbers of photons are being exchanged between the particles in the nail and the particles in the magnet. It is the net effect of all these exchanges that we call the magnetic force. The same thing happens when you comb your hair on a dry day and notice bits of paper sticking to the comb. The force of static electricity is also generated by a flood of photons. At the level of elementary particles, when two electrons collide, we draw the interaction schematically as in Figure 13, two heavy lines indicating the motion of the particles and a wavy line indicating the exchanged photon.

Incidentally, if you had any reservations about the historical unification of electricity and magnetism, they should be dispelled at this point. If both the electrical and magnetic forces arise from the exchange of photons, we should have no hesitation in saying that the two forces are simply different aspects of the same underlying process.

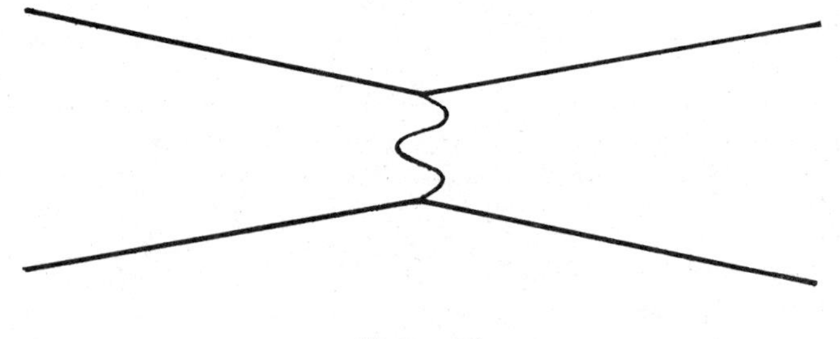

Figure 13.

The zero-mass particle associated with gravity is called the graviton. It is not as familiar to us as the photon, which we see as visible light, and it has not been detected in isolation in the laboratory. We shall see later that there is no clear theoretical understanding of the connection of the subatomic world with the force of gravity, so this state of affairs is perhaps not too surprising. For our purposes, we simply note that the graviton is a massless spin-2 particle whose exchange between objects is thought to generate the gravitational force. If we think of the gravitational force in this way, the old philosophical question about action at a distance simply disappears. There is very definite connection between the earth and the sun—the two are constantly exchanging streams of gravitons. Consequently, it is not surprising that a force can exist between the two. We might also note in passing that the picture of gravity as being caused by particle exchange is very different from the view of gravity presented in general relativity. We will return to this point in Chapter 13.

The weak force has a very short range, so we expect that it will be mediated by heavy particles. According to theory, there should be three such particles. They are supposed to have masses of 80–100 GeV and spin 1. Recalling our classification scheme for elementary particles, any spin-1 particle must be a boson. By historical convention, particles with spin 1 are called vector particles, so the particles that mediate the weak interaction are usually called vector bosons.

There are three vector bosons involved in the weak interactions, differing in the electrical charge they carry. The two charged members of the family are called the W^+ and the W^-, respectively, while the neutral member is called the Z°. One example of the role of vector bosons is seen in the decay of the neutron, a process we have already discussed. The decay actually takes place as shown in Figure 14.

We have already discussed the generation of the strong force via the exchange of pions and other particles, but this is a somewhat superficial way of looking at things. We have argued that the hadrons are not really elementary, and it would seem that the proper way to go about describing the forces between them would be to think of the forces between quarks. Therefore, instead of talking of the exchange of a pion, we would talk of the exchange of a quark and an antiquark that were tightly bound together. In this scheme, the true description of the strong force

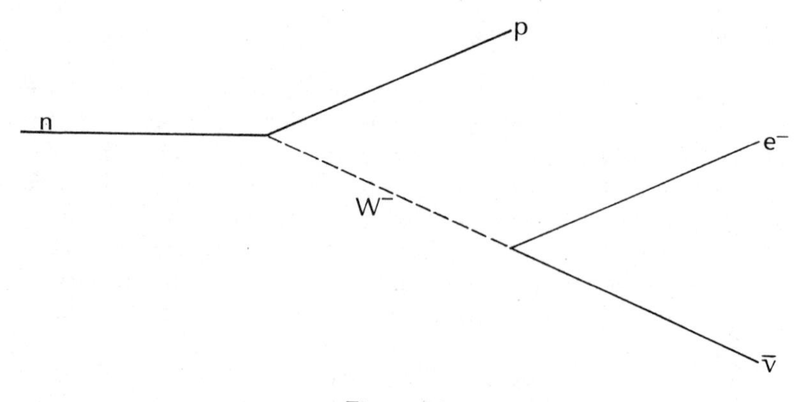

Figure 14.

would concern itself with the particles being exchanged between the quarks.

A full description of the theory of quark interactions will be given later, but we can anticipate some of the results here. It is thought that the force between quarks is generated by the exchange of massless particles called gluons (the name comes from the historical references to the pions as being the nuclear "glue"). These particles carry the color charge between quarks. Although the fact that they are massless might seem to indicate that the force would have an infinite range, the laws that govern color charges are such that it is impossible for an object like the gluon to travel more than about 10^{-13} cm. Thus, the range of the strong interaction remains at the value we have already given it.

A summary of the four forces and the particles that generated them are given in the following table.

Force	*Particle*	*Spin/Mass (GeV)*	*Electrical Charge*	*Other Characteristics*
Gravity	graviton	2/0	0	not yet seen
Weak	vector boson	1/80	±1,0	
Electro-magnetic	photon	1/0	0	
Strong (hadrons)	hadrons	any/.105 or more	any	
Strong (quarks)	gluons	1/0	0	color charge

Summary

There are four fundamental forces known in nature. The forces of gravity and electromagnetism are familiar in daily life. The other two forces operate at the subatomic level and are not immediately available to our senses. They are the strong force, responsible for binding the parts of a nucleus together, and the weak force, which governs radioactive decay. A hierarchy of these forces in terms of their strengths and the distance over which they can be exerted (their "range") is given in the table on page 75.

The historical trend has been to see apparently different forces as nothing more than different aspects of the same underlying mechanism. Newton, for example, showed that the force of gravity operating on earth and the force that moved the planets were identical. In modern terminology, we say that he "unified" these two forces and produced a "unified theory" of gravitation in the process. In a similar way, the connections between the forces of electricity and magnetism led to a unified theory of electromagnetism in the nineteenth century. A major goal of science is to produce a truly unified theory in which all forces are seen to be fundamentally the same.

Modern physicists think of the fundamental forces in terms of particle exchange. If you think of two ice skaters throwing snowballs at each other as they pass, you will realize that the exchange can result in a deflection in the skaters' paths—a deflection we could interpret as arising from the action of a force. At the subatomic level, the same sort of process occurs, with each of the fundamental forces being generated by the exchange of a different sort of particle.

The law of conservation of energy would seem to exclude this sort of exchange, since it would require that the mass energy of the exchanged particle be created out of nothing. The uncertainty principle, however, tells us that the creation of such a particle can take place provided that the particle in question is reabsorbed in a short enough time so that no measurement could detect the violation of energy conservation. This requirement means that the more massive the exchanged particle, the shorter the time it can exist and hence the shorter the range of the force. The force between the protons and neutrons in the nucleus can be thought of as being generated by the exchange of pi-mesons and other hadrons. The weak interaction is generated by a family of very massive particles called vector bosons. With masses around 80 GeV, members of this family carry positive, neutral, and negative electrical charges.

Electromagnetism is generated by the exchange of photons, which explains why the photon was in a category all by itself when we classified particles in Chapter 4. Gravity is thought to be mediated by the exchange of a massless spin-2 particle called the graviton. The fact that the photon and graviton have zero mass explains the long-range nature of these two forces.

Finally, the fundamental strong interaction—the interaction that takes place between quarks rather than between hadrons—is mediated by the exchange of a family of massless particles known as gluons. These particles carry the color charge, and the laws governing color guarantee that this interaction will have a short range, even though the exchanged particle has zero mass.

A summary of the fundamental forces and the particles whose exchange generates them is given on page 84.

Chapter 6

The First Modern Unification

Cannon to the right of them,
Cannon to the left of them, . . .
ALFRED, LORD TENNYSON
"The Charge of the Light Brigade"

The Role of Symmetry in Physics

Modern physicists don't always think in conventional ways. One of the great changes that has occurred in the twentieth century is the increasingly important role that has been assigned to the concept of symmetry. From the study of solids to the development of a unified field theory, the language of symmetry is used everywhere.

There are many symmetrical structures in nature—a snowflake is a good example. The symmetry of a snowflake is apparent to everyone, but what exactly is it? The best way to answer this question is to notice that if I take a snowflake and rotate it through 60° (or 120° or 180°), the result is indistinguishable from the snowflake in its original orientation. We say that the shape of the snowflake is invariant under rotations through multiples of 60°.

The proper way to describe such a symmetry in mathematical terms is to use a discipline called group theory. A group is a mathematical concept but for our purposes we need to know only that group theory has developed a rather arcane language for describing symmetries. We have said that the snowflake can be rotated through angles of 60° without changing its appearance. A mathematician would render this statement as "The snowflake is invariant under the group C_6." The statements are completely equivalent, but the mathematical phrasing provides a more general and more concise way of talking about symmetries.

Another somewhat less intuitively obvious symmetry is that associated with a large spherical tank of water or air. An observer in the center of such a tank would see exactly the same thing no matter which way he turned. We would say that the situation is invariant to any rotation in three dimensions, since up, down, and sideways are all completely equivalent as far as the observer is concerned. The group involved is called $O(3)$, the group of rotations in three dimensions. What goes against our intuition is the fact that the tank of water or air represents a much higher level of symmetry than the snowflake. To understand this comment, remember that the tank will look the same if it is rotated through an arbitrary angle (say 47°), but the snowflake will not.

The central aspect of the role of symmetry in physics can be stated in this way: A symmetry exists if making a certain type of change produces no measurable effect on the system being studied. Rotating the snowflake changes the snowflake, of course, but it does not affect anything about the snowflake that we could measure. That is why it was so obvious that the snowflake was symmetric.

So far we have talked about symmetry in terms of structure. It is, however, a very general concept and can be applied to other things as well. We could, for example, talk about the symmetry of a force. The force of gravity exerted by the sun does not depend on the direction to the attracted object but only on the distance between that object and the sun. If we moved the body to another location but kept its distance from the sun constant, the gravitational force would be unchanged. You can imagine carrying out this operation by attaching the body to a huge sphere centered on the sun and then rotating the sphere. Since no rotation will change the gravitational force, we say that gravity is invariant under rotations in three dimensions.

A magnet does not share this symmetry. The force exerted by a magnet on an object located above its north pole is different from what it would be if the object were located above its south pole. The force exerted by a magnet, then, is not invariant under rotations in three dimensions.

The connection between symmetries involving structure and those involving forces is often quite subtle. For example, the atoms of oxygen and hydrogen that make up the snowflake would be completely symmetric if we looked at them individually. Similarly, the electrical force that governs the behavior of atoms is, like gravity, symmetric under rotation. Yet when two atoms of hydrogen and an atom of oxygen are brought together in oxidation, the electrons of each atoms shift in response to the presence of the others, and the net result is that the water molecule is formed as shown in Figure 15. This molecule is definitely not symmetric under an arbitrary rotation in three dimensions, even though each of its components and the force that governs their behavior is. It is, of course, the 120° angle in the structure of the water molecule that ultimately determines the shape and degree of symmetry of the snowflake.

We say that the underlying symmetry of the atoms is broken when the atoms are assembled into a snowflake. Because this process occurs without any outside interference, we say the symmetry is spontaneously broken. Spontaneous symmetry-breaking is an extremely important concept in the development of the unified field theory because it shows that the interactions that govern a particular phenomenon may actually be much more symmetric and more simple than the phenomenon itself.

At the risk of belaboring this point, let me give another example of spontaneous symmetry-breaking. The interaction of one iron atom with another does not single out any particular direction in space. It is the same no matter how we orient the atoms. The interaction, in other words, is invariant under the group $O(3)$. At high temperatures, this symmetry of the interaction is reflected in the arrangement of the atoms. Sitting inside a large collection of iron atoms, we would see the same situation in every direction—a collection of atoms with their atomic magnets pointing in random directions. (The iron atom, like most others, can be thought of as a tiny magnet with a north and a south pole.) If the iron were cooled, however, this situation would change. At some

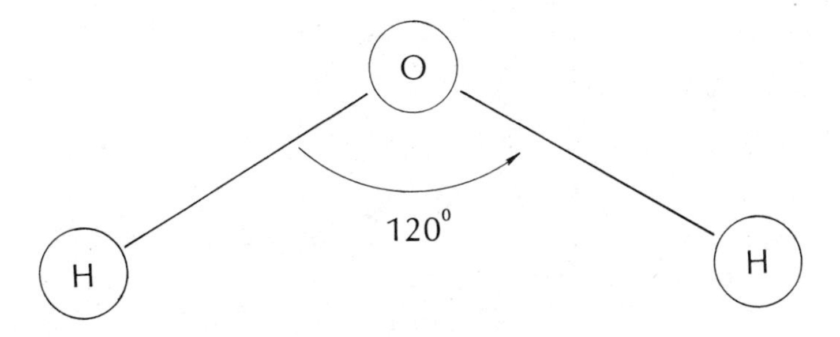

Figure **15.**

particular temperature, the forces acting between the atoms would cause them to line up. The north poles of the atomic magnets would all point in the same direction, producing a large magnetic field. This is what happens in an ordinary iron magnet. The point is that the original $O(3)$ symmetry is spontaneously broken when the magnet forms.

Looking at a magnet with all of the atoms aligned, it would be hard to guess that the underlying interaction was actually symmetric under rotation: the magnet surely is not. Rotating the magnet 90°, for example, produces an entirely different magnetic field from the one we started with. There is, however, one way we could see the basic symmetry of the system. If we added energy to the magnet in the form of heat, the temperature would go up to the point where the alignment of the atoms would be broken up and the system would once again be symmetric under rotation. The lesson we learn from this example is that it is sometimes possible to see underlying symmetries at work in a system by observing the system at very high energies. The effects of spontaneous symmetry-breaking disappear and the original symmetry is restored.

Up to this point we have talked only about geometrical types of symmetries—those associated with rotations in space. The symmetry concept as used in physics is much more general than this, however. There are many kinds of symmetries that cannot be expressed simply as rotations but are nonetheless real. For example, if we had two positive electrical charges as shown in Figure 16, they would exert a repulsive force on each other. If we changed both to negative charges, the force would be unaffected. It would still be repulsive and still have the same value as it had previously. If we had started with one charge positive and one negative and performed the reversal of charges, the force would not change either. We conclude that there is a fundamental symme-

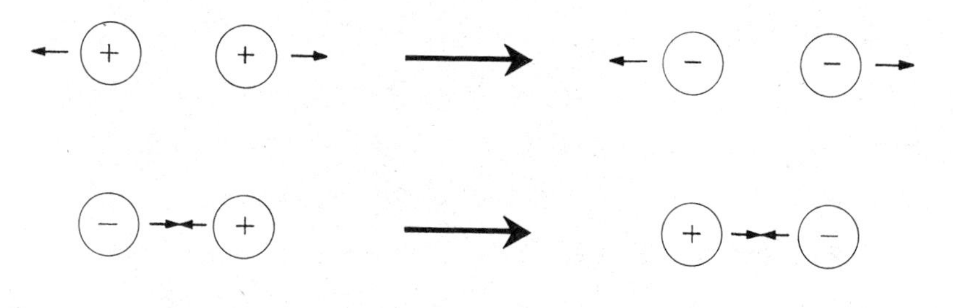

Figure 16.

try in the laws of electricity that says that electrical effects will remain unchanged if we substitute positive charges for negative and vice versa everywhere they appear in the system. This symmetry cannot be represented easily by a rotation in space, but it can be dealt with mathematically in terms of groups.

Another example of abstract symmetry, similar to the symmetry involving electrical charges, is provided by the concept of the strong interaction being produced by the exchange of mesons between particles in the nucleus. If you review the uncertainty principle arguments in Chapter 5, you will see that as far as the strong force is concerned, it makes no difference whether the particle emitting the meson is a proton or a neutron. If we consider only the strong interaction, it is clear that it does not matter if we change protons into neutrons—the force associated with the exchange of mesons will be the same. The "rotation" involved in changing a proton into a neutron is an abstract idea and is not intended to be visualized like the rotation of a ball in ordinary space. Instead, it involves rotations in an abstract space called isotropic spin space. Thus, saying that the strong interactions are invariant if we interchange protons and neutrons is equivalent to saying they are invariant under rotations in isotropic spin space. In the language of group theory, we say they are invariant under the group *SU*(2) in that space.* This is a bit of terminology we will encounter again.

Gauge Theories

A particularly important type of abstract symmetry goes by the name of gauge symmetry. This type of symmetry was conceived of when people were thinking about situations where length scales of a system could be changed. One such situation would be a world located on a rubber sheet, where stretching or compressing the rubber could stretch or shrink the distance between two points. It turned out that some theories were invariant under this sort of transformation, that their predictions did not depend on whether the rubber was stretched or not. Since distance is measured with a gauge, these theories are said to be invariant under gauge transformations and are called gauge theories for short. As

*The letters *SU* stand for *special unitary* and refer to technical properties of the group.

is the case with other types of symmetry we have seen, the concepts associated with gauge transformations can be defined abstractly and applied to systems far removed from a rubber sheet.

Let us start our discussion with a simple experiment. In a laboratory located at sea level, we lift a 1-pound weight 10 feet into the air, measuring the amount of energy expended by our apparatus to accomplish this feat. Next we move the laboratory equipment to a mountain town at an altitude of 5,000 feet and repeat the experiment. For the moment, let us ignore the fact that the real force of gravity decreases slightly as we move away from the center of the earth and assume that it takes the same force to lift the weight at 5,000 feet as it did at sea level. It makes no difference whether the weight is moved between 0 and 10 feet or between 5,000 and 5,010 feet—the amount of energy required is the same. This is illustrated on the left in Figure 17. This result must represent another symmetry in nature.

The same sort of effect can be seen with electrical forces. If we perform an electrical experiment, then repeat the experiment after the entire laboratory has been connected to the pole of a huge battery as shown on the right in Figure 17, no perceivable change will result. Raising the voltage of everything in the laboratory is equivalent to changing the altitude in our previous example. The only important thing is the difference in voltage between points in the apparatus, and not the absolute value of the voltage. You have often seen a striking verification of this statement, although you may not have interpreted it as such. A bird can sit on an exposed power line without being electrocuted, even though the line may be at 67,000 v or more. The reason for this is that even though each of the bird's feet are at 67,000 v, the difference in voltage between the two feet is zero, so no current will flow.

This sort of transformation, which amounts to redefining what is meant by the zero level of height or voltage, is the simplest kind of gauge transformation. It is an example of what physicists call a global symmetry; to make it work, we have to raise everything to the same height or the same voltage. If only half the laboratory were raised to a height of 5,000 feet, for example, there would be clear differences between the systems, as you could quickly discover by trying to walk from one side of the laboratory to the other. The fact that electricity must exhibit this symmetry allows us to rule out a large class of possible theories.

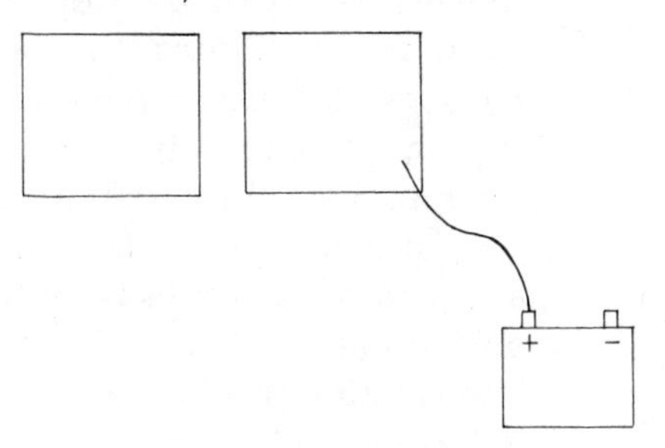

Figure 17.

We would not, for example, accept a theory that stated that the electrical force depended in some way on the absolute voltage at a point rather than on the difference in voltage between two points.

This sort of symmetry actually represents a deep-seated intuition we have about nature. Nothing in nature ought to depend on the state of mind of the person observing it. The elevation I choose to call zero is completely arbitrary and can vary from one observer to the next. The requirement of gauge symmetry is simply the statement that every observer must see that same force of gravity, no matter how he wants to define the zero of height. As we go on to more complex gauge theories, this will be our guideline: Nothing in nature can depend on arbitrary definitions that might vary from one observer to the next.

There is a stronger version of this principle. We have seen that we are free to set the zero of height or voltage arbitrarily but that once it is set at one point in space, we must stick with that definition everywhere. This is called a global gauge symmetry. Suppose, however, that it was possible to imagine a system where we were free to set the zero of voltage arbitrarily at each point in space, regardless of what had been done at a neighboring point. If such a theory could be constructed, it would obey a local gauge symmetry, rather than a global one. Clearly, neither Newtonian gravitation nor ordinary electricity are theories that contain a local symmetry, since both allow only one choice of zero throughout space. Yet, if we consider the philosophical basis of gauge symmetry, we should begin to wonder why the electrical potential at point *A* should depend on the arbitrary definition of zero by someone at point *B*. A theory that possessed a local gauge symmetry would not depend on arbitrary definitions of a single observer, and the definition of one observer would be completely independent of the definition of another.

One way of understanding the difference between global and local symmetries is to think about the meridians of longitude on the earth's surface. By convention, everyone agrees that the line running through the Greenwich Observatory near London is to be assigned the number zero, and all other longitudes are to be measured relative to Greenwich. If this convention were changed and the zero were assigned to some other place (Tokyo or Chicago, for example), there would be no difference in any measurable quantity. The distance between New York and London would still be about 74° of longitude, regardless of what we called zero. The distance between any two points would still be the same number of degrees, and all that would change would be the zero point from which all longitudes are measured. This is an example of a global symmetry.

If, on the other hand, we allowed each person to choose his own zero of longitude, the situation would be more complicated. If we wanted to calculate the difference in longitude between London and New York in this case, we would have to take the number that an observer at each place had assigned to his longitude and subtract. If both observers had insisted on designating their location as zero longitude, our calculation would yield the result that the distance between the two was zero degrees—clearly an incorrect conclusion. Thus, the system of measurement on the earth's surface is an example of something that exhibits global, but not local, gauge symmetry. This shows that finding things that exhibit local gauge symmetry may be easier said than done.

Electricity, like longitude and Newtonian gravity, exhibits a global gauge symmetry but not a local one. The key point in our discussion of electricity and gravity, however, is that changing the zero of height or voltage globally does not alter the electrical or gravitational force exerted on any body. Making the change in definition produces no measurable change in the world. If we think about electricity alone, it is clear that we cannot arbitrarily fix the zero of voltage at two different points without changing the force an electrical charge would feel. Thus, electricity by itself does not exhibit a local gauge symmetry. If, however, we recall the intimate connection between electricity and magnetism, we can ask another question: Is it possible that the changes in the world produced by making a local gauge transformation for electricity could be compensated for by the changes that would be induced by doing the same thing for magnetism? Is it possible, in other words, that electricity and magnetism, neither of which displays local gauge symmetry when considered alone, could conspire in just such a way to produce a theory that does display this symmetry?

It turns out that this is exactly what happens in the classical Maxwell theory of electromagnetism. Changes in the electrical potential produced by an arbitrary change of zero at each point in space can be precisely canceled out by making an appropriate change in the magnetic potential. In other words, no matter how arbitrarily we vary one potential, we can always adjust the other so that no observable consequences follow—no forces are changed, no trajectories altered. This represents an important generalization of the idea of the gauge symmetry as being associated with changes in definitions of zero levels of height or voltage.

Perhaps the most important result to take away from this discussion is that while neither electricity nor magnetism taken alone would exhibit a local gauge symmetry, the unified theory of electromagnetism does, because in that theory the parts of the electrical effects that violate the symmetry are canceled out by magnetic effects and vice versa. It appears that unification is not only intellectually beautiful but gives us symmetric theories in the bargain. For the sake of completeness, I should say that the fact that electromagnetism is invariant under combined electrical and magnetic changes of the type we have been discussing is rendered in the language of mathematics as "The theory is invariant under $U(1)$ gauge transformations."

Another way of looking at gauge symmetry is to note that the only thing we can actually see in an experiment is a change in the state of motion of some quantity of matter. In other words, we can only detect the presence of forces when they act, nothing else. If there is some way in which we can change the system without changing the forces, a situation we encountered with our fictional laboratory, that change in the system will be undetectable by any experiment. Hence, it can make no difference in nature whether the change occurred or not. Our theories must be such that they reflect this fact.

In the case of electromagnetism, we know that the description of gauge symmetry we have just given must be reflected in some way in the exchange of photons between charged particles. In other words, quantum mechanics, the science devoted to describing the behavior of elementary particles, must also contain some version of gauge symmetry. The way that this version arises is particularly interesting. It turns out that making the kinds of changes we have been describing in electric and magnetic potentials requires a corresponding change in the description of a charged particle we get from quantum mechanics. If we want quantum mechanics to have the same kind of local gauge symmetry that classical electromagnetism has, then we must demand that there be no change whatsoever in any measurable quantity when observers at different points change their definition of voltage independently of each other. In the classical case, we found that the way to produce this kind of local symmetry was to have two effects that canceled each other out.

In the case of a force being generated by an exchange, there are two things that could be changed. One, already mentioned, is the quantum mechanical description of the particles on which the force acts. The other, which we mention now for the first time, is the quantum mechanical description of the particle being exchanged. We can pose the question of local gauge symmetry in a theory describing particles as follows: Is there any way that the changes in the descriptions of the charged particles and the particle being exchanged could cancel each other out, leaving us with a theory that exhibited local gauge symmetry?

The answer is astonishing. It turns out that such a cancellation is possible, but only if the particle being exchanged has zero mass and spin 1. In this case, and in this case only, will the delicate interplay that we saw in the classical case repeat itself at the level of the particle. The particle with spin 1 and zero mass is, of course, the photon. So beautiful is this result that theoretical physicists often speak of the very existence of the photon as being the result of the requirements of gauge symmetry.

The Electroweak Unification

We now have at our disposal all the concepts we need to understand a chain of events that began in the late 1950s and culminated in the awarding of the Nobel Prize in physics to three theoretical physicists in 1979. Until recently, gauge theories were the preserve of a few theoreticians far from the main currents of high-energy physics. To give just one bit of evidence for this claim, during my own years of graduate study at Stanford in the mid-1960s, I heard only a few offhand references to gauge theories and read only one paper on the topic, more for general background than anything else. The popular topics of that period, which carried names like *Regge poles* and *duality*, have, like the hula hoop, gone their way to be replaced by theories that were once neglected.

In the 1950s an undercurrent of thought in theoretical physics was that there is probably some sort of deep connection between the electromagnetic and the weak interactions, primarily because both involve the exchange of spin-1 particles. It was known that it is possible to put together (or "construct," as theorists say) a theory that had local *SU*(2) gauge symmetry. What this meant was that you could put together a theory in which it was possible to start at one point in space and change a neutron into a proton (or vice versa), and then go over to the next point in space and do the same thing without any reference to what you had done at the first point.

This involves a rather large extension of the idea of gauge symmetries. We started with simple gauge transformation, which amounted to little more than a change of our choice of zero. We moved up to a more complicated symmetry that required an interplay between electric and magnetic potentials but still depended ultimately on the arbitrary definition of zero points.

The next step in complexity involves a more abstract kind of symmetry, one that is somewhat more difficult to visualize. In Chapter 5 we saw that the strong force between protons and neutrons could be thought of as being generated by the exchange of a meson. In such an exchange, it makes little difference whether the particles involved are protons or neutrons—the strong force will be the same in either case. This situation has led physicists to speak of isotropic symmetry in nature, a symmetry in which one imagines that it is possible to redefine electrical charge in much the same way we redefined the zero of height or of voltage. If this symmetry were observed in nature, then it would make no difference if we went into a nucleus and changed all of the protons into neutrons and all of the neutrons into protons. Since it actually makes very little difference if this is done in real nuclei, physicists believe that this operation represents a symmetry in the same sense that redefining zero height does.

One way of thinking about isotropic symmetry is to imagine a dial at each point in space. If the dial points up, the particle at that point is a proton; if it points down, the particle is a neutron. Interchanging protons and neutrons inside a nucleus, therefore, corresponds to turning each dial 180°. Where there was a proton, there is now a neutron and vice versa. If the world is unchanged by this operation, which amounts to a redefinition of electrical charge everywhere in space, we would say that nature is invariant under a global isotopic spin symmetry. In the jargon of mathematics, such a theory exhibits global *SU*(2) gauge symmetry.

In the light of our previous discussion, the next question is obvious. Can there be a theory where the isotropic symmetry is local rather than global? In other words, could we produce a theory in which going through space twisting the dials at random at each point produced no measurable effect? Our first reaction to this question would probably be that such a situation was not possible. It turns out, however, that this first reaction is wrong. Just as it was possible to produce a local symmetry in electromagnetism by canceling the changes associated with charged particles by the changes associated with the exchanged photon, it is possible to put together a theory in which protons and neutrons can be interchanged at will at each point in space, and to have the changes associated with this interchange canceled by changes in the particles being exchanged to generate the force. It turns out that a theory involving protons and neutrons will possess local *SU*(2) gauge symmetry if the forces are mediated by a family of four massless spin-1 particles (some of them carrying electrical charge) being exchanged to generate the force. In this case, as was the case with electromagnetism, the changes in the description of the particles caused by the transformation is exactly canceled by the changes in the exchanged objects, and everything in the theory is the same as it was originally. Thus, for a force generated by this family of particles, changing protons to neutrons and making the accompanying changes in the exchanged particles has no more effect on the system than moving our lab from sea level to 5,000 feet did in our previous example. The predictions of the theory are invariant.

This fact was well known among physicists for a long time (in fact, the classic exposition of the idea was the lone gauge theory paper I read in graduate school). It remained as little more than a curiosity, however, because it required that there be four particles like the photon, two of which were supposed to be charged. No such particles had ever been seen in the laboratory, so it was felt that the theory, while intellectually pleasing, could have nothing to do with the real world.

To the extent that the theory could be taken seriously, however, physicists felt that it had to have something to do with both the electromagnetic and the weak interactions. Both of these interactions are mediated by the exchange of spin-1 particles (the photons and the vector bosons) and could therefore be expected to have some similarities. The fact that the *W* bosons are supposed to be massive, while the particles predicted by the theory were supposed to be massless, remained the sticking point as far as any application of the theory was concerned.

In 1967, Steven Weinberg, then at the Massachusetts Institute of Technology, published a paper that was to revolutionize physics. In essence, he showed that the gauge theories of the type we have discussed could describe the real world if the effects of spontaneous symmetry-breaking were taken into account. In one of those events that seem too contrived even to appear in fiction and that therefore happen only in real life, the same discovery was made independently a few months later by Abdus Salam in London.

We have seen that a group of iron atoms will align themselves in a given direction at low temperatures, even though the interaction between atoms has no preferred direction in space. In so doing, the atoms acquire a certain energy, and we have to add energy to the system (for example, in the form of heat) to break the alignment and see the true symmetry. In the case of the weak interactions, the theory predicts that the underlying symmetry is such that the four particles being exchanged to generate the force must be massless. What Weinberg and Salam showed was that at low energies a spontaneous symmetry-breaking occurs and three of the exchanged particles acquire a mass, while the fourth remains massless. This is analogous to the case of the magnet, where the alignment of the atoms gives the system an energy it would not have if the symmetry were not broken. In the case of the particles, this added energy takes the form of a mass for the exchanged particles.

With this development, one of the major objections to the gauge theory was answered. The theory no more predicts the existence of four massless spin-1 particles at normal energies and temperatures than the theory of the behavior of iron atoms predicts that there will be no preferred direction in space at room temperatures. In both cases, the true symmetry in nature is masked by the process of spontaneous symmetry-breaking. We can deduce what that true symmetry must be only with the insight provided by our theory. The spatial symmetry of the iron is broken spontaneously when the temperature falls to the point where the magnetic fields line themselves up in a given direction. In a process that is the exact analog of the formation of a magnet, the same thing happens with the four particles predicted by the gauge theory. The full symmetry is seen in nature at very high energies when the four particles have masses that are negligible compared to the energies of collision. As the temperature drops, however, this symmetry is spontaneously broken. Just as a certain amount of energy is tied up in the creation of large-scale magnetic fields in a magnet, a certain amount of energy is tied up in the four particles when the symmetry is broken. Instead of the energy going into a magnetic field, however, in the case of the particles, it goes into mass. It turns out that if there really was an underlying local $SU(2)$ gauge symmetry in nature, the effects of spontaneous symmetry-breaking would be such that in our present era we would not expect to see four massless particles in the laboratory. The theory predicts that three of these four particles would acquire a mass when the symmetry is broken, by a process analogous to the way the magnetic field acquires energy when the magnet is formed. The fourth particle, for technical reasons, remains massless. Thus, the theory predicts that we should see one massless spin-1 particle in nature (which we can identify with the photon) and three massive spin-1 particles. Two of these massive particles should carry electrical charge, and we can identify them with the ordinary charged vector bosons. The third heavy particle is electrically neutral and represents a new kind of particle involved in the weak interactions, a vector boson given the name Z°. The masses of the bosons were supposed to be 80–100 GeV.

Just as we argued that electricity and magnetism were merely different aspects of a single force because they both involved the exchange of the same particle (the photon), the Weinberg-Salam theory shows that electromagnetism and the weak interaction are the same in that they arise from the exchange of the same family of particles. The fact that the interactions appear to be so different has to do with spontaneous symmetry-breaking and does no more to negate this conclusion than the aligning of iron atoms proves that the underlying atomic force depends on direction. There are some symmetries in nature that are hidden from direct view and that can only be seen at high energies. The basic symmetry of the atomic force becomes apparent when we heat the magnet. The basic symmetry of the electromagnetic and weak forces should become apparent when enough energy is available in collisions to make the differences in masses of the photon and vector particles irrelevant. In practice, this means we must look at energies larger than the 80–100 GeV mass of the particles. In terms of the ages of the universe outlined in Chapter 2, the temperature at the start of the particle era, a fraction of a millisecond after the Big Bang, corresponded to only a few giga-electron volts in collision energy. Even at that point, the temperature was too low to see this symmetry, and it is certainly too low in our present, relatively frigid era.

The end result of the Weinberg-Salam theory, therefore, is that it is no longer necessary to think of the weak and electromagnetic forces as being distinct and separate. Because we now understand that they are associated with the exchange of the same family of particles and that the apparent differences between them are the result of spontaneous symmetry-breaking, we can reduce the number of fundamental forces from four to three. The new force, mediated by the exchange of spin-1 particles, we call the electroweak interaction. We recognize that this unification is a hidden symmetry in today's world but would become manifest in a world where the energy associated with ordinary collisions would greatly exceed the 80–100 GeV mass of the vector particles.

Evidence for the Electroweak Unification

After Weinberg and Salam published their papers, there was a period of several years when they were largely ignored. For example, there is a publication called the *Science Citation Index*, which keeps track of the number of times researchers refer to a given publication. In the five years between 1967 and 1971, Weinberg's paper was cited a total of five times (compared to hundreds per year in the period that followed). Starting in 1971, however, a series of verifications of the unified gauge theory began to arrive on the scene. Some of these were experimental and some theoretical.

THEORETICAL EVIDENCE

We have discussed the connection between exchanged particles and forces in terms of diagrams like that on the left in Figure 18, where a single particle is exchanged. It is obvious, however, that this is not the only process that can occur. In the case of electromagnetism, it is clear that there are many other ways that two charged particles can interact. For example, one particle could emit a photon, reabsorb it, and then exchange a photon with the other particle. This process is shown in the center of Figure 18. Other, more complicated types of interactions can occur, one of which is shown in Figure 18 on the right. Since we cannot, because of the uncertainty principle, tell which of these processes actually happens, we calculate the electromagnetic force by adding up the contributions from all of the different sorts of diagrams and say that the actual force we see in the laboratory is the sum of all possible ways the interaction could happen. The branch of physics devoted to the study of these sorts of calculations is called quantum electrodynamics (QED) and contains the most impressive theory we have. It predicts many experimentally measurable quantities to accuracies of one part in a billion or better—an accuracy unequaled anywhere else in science.

There is one problem with this idea of adding up contributions from different sorts of processes. As soon as we get away from the exchange of a single particle, we find that the probabilities for some of these interactions become infinite. Even the cen-

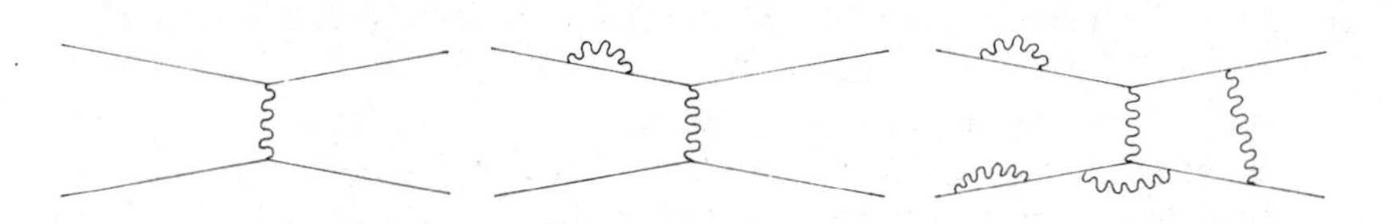

Figure 18.

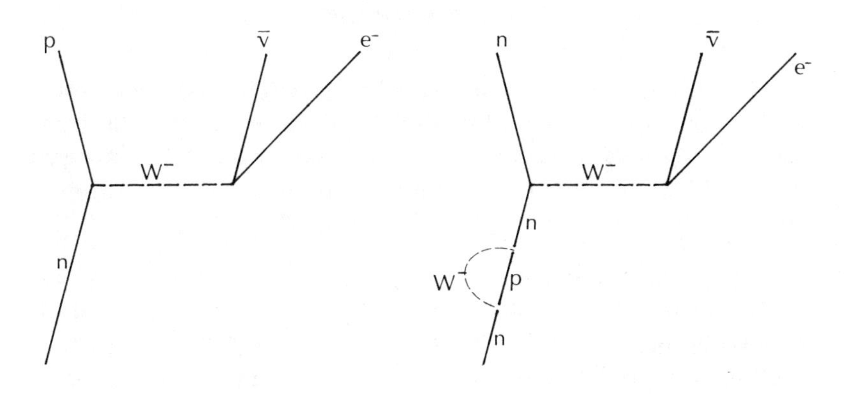

Figure 19. *Figure 20.*

ter diagram in Figure 18 has this characteristic. One of the great achievements of the late 1940s was to show that for every positive infinity that comes out of our mathematical calculation, we can find a negative infinity that cancels it. The net result is that the predictions of QED for anything that can be measured are perfectly finite, and there are no problems. The infinities are caused by the way we do the mathematics, not any physical effect. Theories in which infinities can be canceled out in this way are said to be renormalizable. QED is an example of such a theory.

The original theory of the weak interactions did not have this property. Because the weak interactions that had been observed before 1967 were all similar to the decay of the neutron, in the sense that they involved the exchange of charged vector bosons, it was believed that there were only two massive particles mediating the weak interaction, the W^+ and the W^-. The decay of the neutron took place as shown in Figure 19. Taking our lead from QED, we know that there are other ways for the neutron to decay (one is shown in Figure 20). As with QED, the more complicated graphs give rise to infinities, but unlike QED, these infinities cannot be canceled neatly. As long as we include only the charged vector particles in our calculations, the theory of weak interactions is not renormalizable. This means that the theory must predict that some quantities involved in weak interactions are infinite, a result that is simply nonsensical. There are no infinities in the laboratory.

The gauge theories, however, predict that there should be a third massive particle involved in the weak interactions—the Z°.

In 1971 the Dutch physicist Gerard t'Hooft (then a graduate student) showed that the addition of the Z° provided just the right kind of infinities to cancel those generated by the W^{+} and the W^{-}, and that the theory including all three was renormalizable. In the words of one commentator, this development "revealed that Weinberg and Salam's frog was an enchanted prince." Far from being an added complication, the Z° provided just the piece of the puzzle that had been missing, solving an old and very difficult theoretical problem in the process. It was at this point that high-energy physicists began to jump on the gauge theory bandwagon.

EXPERIMENTAL EVIDENCE

The prediction of a third vector meson has important consequences for experimental physicists. The first to be investigated involved interactions initiated by neutrinos. It is possible to make beams of neutrinos in high-energy accelerators by creating a beam of particles, such as pi- or mu-mesons, whose decay products include neutrinos and then to direct this beam into a long pile of dirt. Eventually the initial particles decay, and the dirt absorbs everything except the neutrinos. The result is a beam of pure neutrinos whose interactions with selected targets can be studied.

If a neutrino strikes a proton and if only charged vector bosons exist, then the neutrino must change its identity. A sample of this sort of interaction is shown in Figure 21. In interactions involving the exchange of charged vector mesons, there will always be a mu-meson or electron produced along with the debris of the proton. If, however, there is a Z°, we can have an interaction like the one in Figure 22, where the debris is not accompanied by a charged lepton. In 1973, groups working at the Fermi National Accelerator Laboratory near Chicago discovered such events. Their work was soon confirmed by experimenters at the European Center for Nuclear Research in Geneva. The Z° particle predicted by the gauge theories was seen (albeit indirectly) in experiment, and an important piece of evidence in favor of the theory had been obtained.

The theory also made predictions about a large number of other processes involving weak interaction processes, all of which depended on one number that could be calculated from the the-

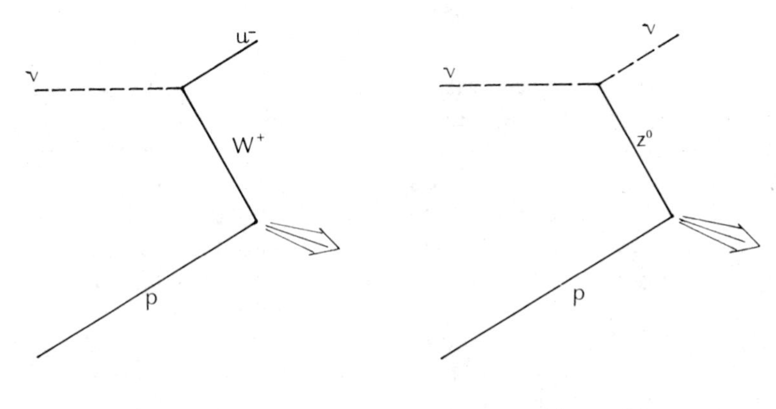

Figure 21. *Figure 22.*

ory. For various technical and historical reasons, this number was expressed as an angle known as the Weinberg angle. The theory gave this angle a value of about 22°, and on the basis of this number all sorts of predictions about processes ranging from high-energy interactions to the emission of radiation by atoms could be made. During the period between 1973 and 1979, battles raged between experiment and theory over whether these predictions were correct. Experiments were done that seemed to contradict the theory, the theory would be patched up to agree with experiment, the experiment would be redone and new results obtained, the theory was repatched, and so on. When all of the dust had settled and all of the experimental results agreed upon, the predictions of the original theory were confirmed.

Thus, by 1982, only one important experimental confirmation of the theory remained to be carried out, and that was the isolation and identification of the vector mesons themselves. There was, as we have seen, indirect evidence for the existence of the neutral particle, and physicists were confident that all three will be seen in the laboratory. In a sense, finding the vector mesons was not so much a test of the gauge theory as it was a test of the idea that the weak interaction is mediated by the exchange of a particle. Every theory of weak interactions predicts the existence of the W^+ and W^-; the only new element from the gauge theories is the addition of the $Z°$. The confirmation of this fundamental idea was delayed for a long time for a very simple reason. If the masses of the particles are really in the range 80–100 GeV, as current theories suggest, then it requires a great deal of energy to create them in accelerators. Since all of the energy that goes into the masses of these particles must come from the energy of accelerated protons, accelerators must be very powerful indeed if we are to have any chance at all of seeing the vector mesons.

In fact, it was not until the spring of 1981 that a machine capable of producing the vector particles was put into operation at the European Center for Nuclear Research. In this machine, beams of protons and antiprotons, each with an energy of 540 GeV, collide head on. Some of the possible reactions that might result from these collisions are shown in Figure 23. In each case, a vector meson is produced, lives its short life, and decays. The particle cannot be seen directly, because its lifetime is too short. The experimenter's job is to search for electrons or muons of the right energy to have been produced by the decay of the vector meson.

On January 13, 1983, at a workshop in Rome, Carlo Rubbia of Harvard University announced the first evidence for the discovery of a vector boson. He was the head of a group working at the European Center for Nuclear Research in Geneva. Rubbia's group found six events in which electrons such as those expected from the diagram on the right in Figure 23 were seen. In each of these events, as expected, their detectors failed to record the amount of energy and momentum that would have had to be carried by the neutrino that accompanies the electron. (The apparatus was such that one would not see the neutrino at all.) The mass of the charged W boson deduced from the six events was 81 ± 5 GeV. This is to be compared to a theoretical prediction of 82.1 ±2.4 GeV. Furthermore, the theory predicts that the group should have seen 4.9 events of the type they detected. On the basis of these comparisons, it is safe to say that the vector boson has now been seen in the laboratory.

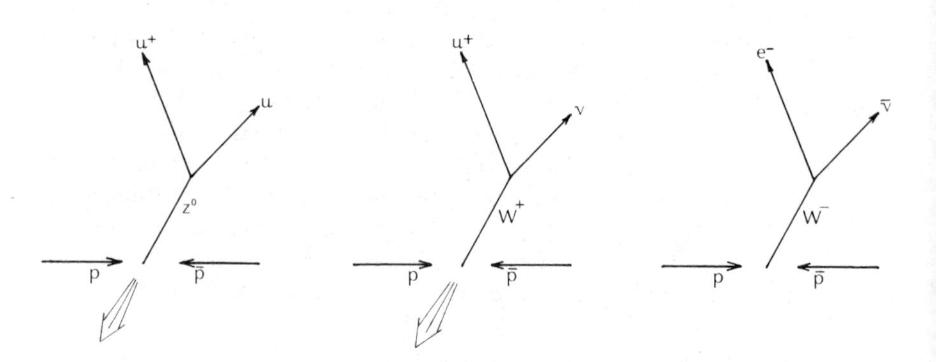

Figure 23.

A Simple Picture of Unification

Because the unification of forces plays an extremely important role in the early evolution of the universe, it is useful to have a simple picture of the unification process in mind. On page 76 we discussed the idea of force as an exchange of particles in terms of an analogy of two skaters throwing something at each other in passing. Suppose that we look at this analogy again, but now imagine two sets of skaters. In one pair, one partner throws a bucket of water at the other, leading to the appearance of a force. In the other pair, a bucket of liquid is also exchanged, but in this case the liquid is alcohol instead of water.

So long as the external temperature is above 32°F, the two forces acting will appear to be much the same. The deflection caused by the two liquids will not be too different, and we would not be uncomfortable in saying that the two forces were identical. When the temperature falls below 32°F, however, the water will freeze while the alcohol will not. Now one set of skaters is exchanging liquid alcohol, the other a block of ice. The force would appear to be very different in this case. In this example, we find two forces that appear very different at low temperatures revealed to be identical at high temperatures. The unification of the electromagnetic and weak forces can be thought of in much the same way.

Summary

In modern physics the idea of symmetry plays an extremely important role. A symmetry exists in nature if making certain types of changes does not affect any measurable quantity in the system being studied. An easy symmetry to visualize is that involved in rotating a snowflake through 60°.

Symmetries can be subtle. The force that operates between two atoms of iron, for example, does not single out a special direction in space, but when iron atoms form a magnet, there is a very definite "north." The interaction between the atoms represents a hidden symmetry in this system, one that is not obvious to the observer. When an interaction produces an effect like that in a magnet, we say the underlying symmetry is spontaneously broken.

So-called gauge symmetries are important in modern physics.

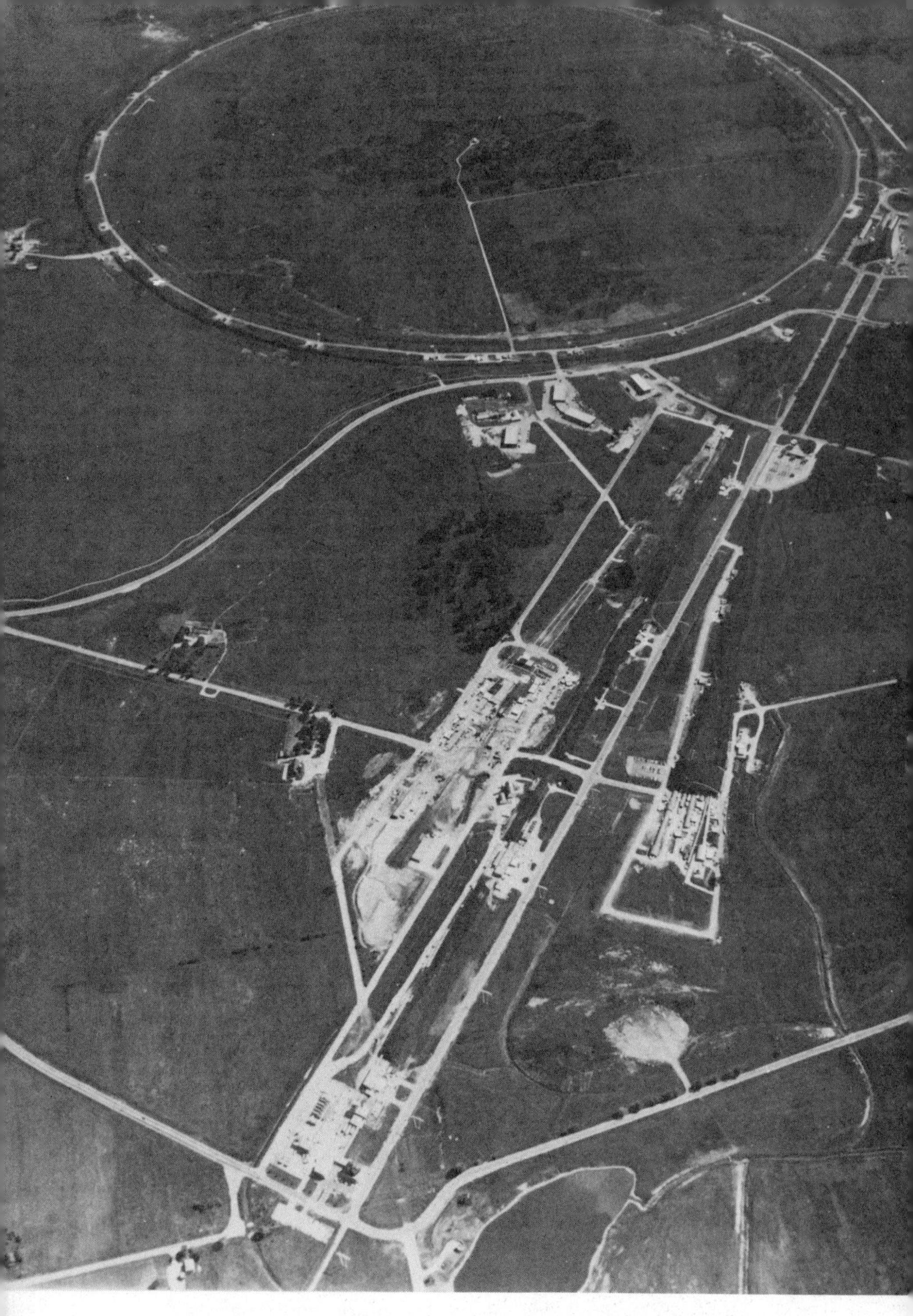

An aerial view of the Fermi National Accelerator Laboratory at Batavia, Illinois. The largest circle is the main accelerator, four miles in circumference. Three experimental areas extend at a tangent from the accelerator. Fermilab photo.

A portion of the equipment of Experiment #1A in which the neutral current was first seen at Fermilab. Fermilab photo.

Photo taken during the vector meson experiment at the Center for Nuclear Research in Geneva. The tracks mark the paths taken by elementary particles in the apparatus. Photo courtesy of CERN.

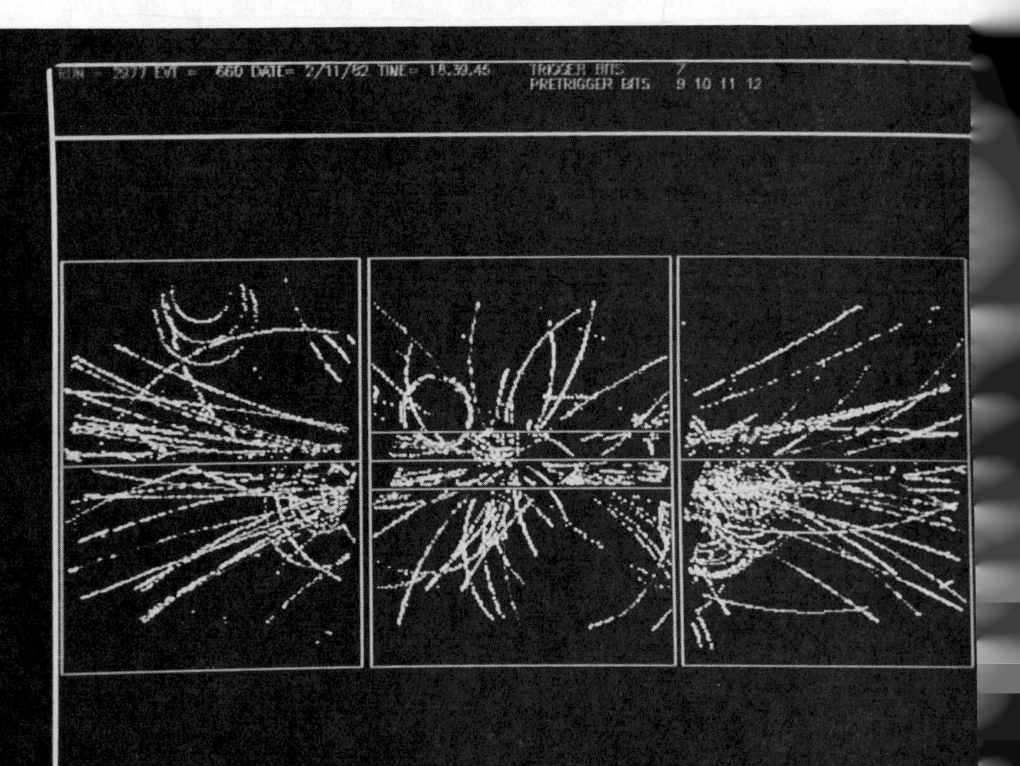

In electromagnetism, it is possible to change the electrical potential (the voltage) of a system in a completely arbitrary way at each point in space without altering any measurable quantity, provided that at each point the magnetic potential is altered to compensate. In the jargon of mathematics, electromagnetism is "invariant under a U*(1) gauge transformation." At the level of elementary particles, this property of electromagnetism operates in this way: Changes in the description of the charged particles due to changes in potentials are exactly compensated for by changes in the description of the photon that is being exchanged.*

In 1967 an important advance was made. A theory was published in which the gauge transformation took the form of the ability to transform protons into neutrons, and vice versa, completely arbitrarily at each point in space. As with electricity, the effects due to this change can be compensated for by changes in the particles being exchanged. The underlying symmetry is such that there must be four massless particles exchanged to generate the force, but the effects of spontaneous symmetry-breaking result in three of those particles acquiring a mass in the range 80–100 GeV, while the fourth remains massless. The three heavy particles are identified with the vector bosons that mediate the weak interaction, while the massless particle is identified with the photon. This theory unifies the electromagnetic and weak forces, since both are now seen as the result of exchanges of the same family of particles. The fact that the two forces appear to be different at present is a result of the fact that there is not enough energy available in collisions to allow the underlying symmetry to be seen. When the temperature is such that energies of 80–100 GeV are available, the two forces will be identical. In any case, this development reduces the number of fundamental forces from four to three, the combined force being given the name electroweak. *Since 1971 a great deal of evidence, both theoretical and experimental, has accumulated to confirm the notion of the unification of these two forces.*

An easy way to visualize the unification of two forces is given on page 107.

The Grand Unification

A scientist is a man who can find out anything, and nobody in the world has any way of proving whether he found it out or not, and the more things he can think of that nobody can find out about, why the bigger the scientist he is.

WILL ROGERS

Quantum Chromodynamics

The most important lesson that we can derive from the marvelous success of the unification of the electromagnetic and weak interactions is that the intuition that told us that there was something wrong with a universe governed by four different theories was correct. We have seen that the idea of hidden symmetries in nature can be invoked to reduce this number to three, and it is natural to ask whether the same approach might not suffice to reduce the number to two or even to one. This possibility did not escape theorists in the early 1970s, of course, and enormous effort was put into producing a further unification through the use of gauge techniques. The effort was successful, and we now have a theory in which the strong force is unified with the electroweak. We will see how the theory works shortly, but before we can speak of unifying the theory of strong forces with anything, we have to understand what that theory is.

We have already seen that the fundamental strong force acts between quarks and that, in addition to their ordinary electrical charge, these quarks carry a type of charge we have termed color. The theory that describes the interaction involving the electrical charges of particles via the exchange of a photon is called quantum electrodynamics. The term *quantum* tells us we are dealing with particles, and the term *electrodynamics* tells us we are looking at electromagnetic effects. By analogy, the theory that has been developed to describe the strong interaction—an interaction involving the color charge of the quarks—is called quantum chromodynamics (QCD), with chromo- referring to color. It is a theory that is similar to, but more complicated than, ordinary electromagnetism. As we shall see, this similarity facilitates the eventual unification of the strong force with the electroweak.

The idea that particles can carry an electrical charge is familiar to us. There are two types of electrical charge, which we call positive and negative. In the normal course of affairs, particles form themselves into atoms in such a way that each atom contains as many positive charges as negative ones. Another way of stating this fact is to note that the laws that govern electrical charge are such that charged particles group themselves together into structures that are electrically neutral.

A similar phenomenon occurs when we examine the color charge on quarks. Instead of two kinds of charge, there are three, which we will call red, blue, and green. We have to remind ourselves that the term *color* here is not used in its ordinary sense. Saying that a particular quark is green is just a shorthand way of saying that the quark carries the type of charge we have arbitrarily labeled green; it is analogous to saying that the quark is positive when we mean it carries a positive electrical charge.

The basic reason why electrically neutral atoms form from protons and electrons is that the law of electrical force is such that like charges repel each other and unlike charges attract. As you might expect, the laws that govern the color charge are somewhat more complex because of the existence of three types of charge rather than two. It appears, however, that there are only two situations in which the color force is attractive. The force between a quark carrying a given color and the antiquark carrying the anticolor is attractive, and the force between three quarks, no two of which have the same color, is also attractive. Every other combination leads to a repulsive force.

Recalling the quark structure of the elementary particles, we realize that the first situation (quark plus antiquark) is what we call a meson, while the second (three quarks) corresponds to a baryon. Thus, the fact that the only kind of strongly interacting particles are mesons and baryons is reflected in terms of color charge by this force rule.

The term *color,* while it does not refer to any pigmentlike attribute of the particles, nevertheless provides a useful mnemonic for thinking about this type of charge. There are three types of color charge on the quarks, and there are three primary pigmentations. Any color of paint can be made by mixing together blue, green, and red in appropriate proportions. If we mix equal amounts of these three pigments we get white—the absence of color. Similarly, there are three pigments that are said to subtract, because they absorb the primary colors when added to a mixture. These are cyan (a blue-green color that absorbs red), magenta (which absorbs green), and yellow (which absorbs blue). Adding a pigment to its subtractive counterpart (for example, red to cyan) produces white.

In terms of this analogy (and I cannot stress too strongly that it is just an analogy), the law for the force between quarks becomes very simple. Quarks can only come together in collections where the resulting color is white. If we think of white as being colorless or color-neutral, there is an exact analog between the color and electrical charge, because electrically charged objects also form stable systems if the final combination is electrically neutral. In terms of the quark colors, the positively charged pion and the proton are shown in Figure 24. Note that a given electrical charge will actually appear on three quarks, each with a different color charge. Thus, what we have called the *u*-type quark (electrical charge ⅔) is actually three different particles—a red *u*-quark, a blue *u*-quark, and a green *u*-quark.

We saw that the way to produce a gauge theory of the weak interactions was to consider symmetries involving the altering of electrical charges at different points in space. In dealing with the quarks, the analogous type of symmetry would involve the color charges rather than the electrical ones. Following the procedure outlined by Weinberg and Salam, we demand that no measurable quantity change when we go through a system altering the colors of quarks at random ("I'll leave this one red, change this blue to

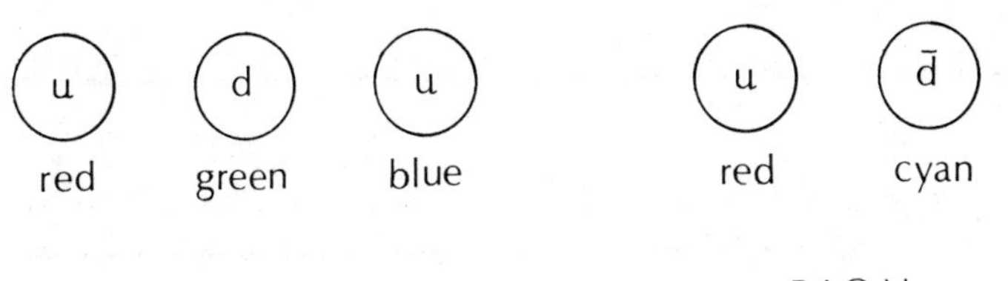

Figure 24.

green, change this green to red, leave this blue alone . . ."). Clearly, if this is all we did, the theory could not be invariant, since changing the color of one quark in a proton would do away with the "whiteness." As was the case with the Weinberg-Salam theory, the changes in the quantum mechanical properties of the quarks caused by this operation must be canceled out by compensating changes in the quantum mechanical properties of the particles being exchanged between the quarks, particles we have called gluons. In order for such a cancellation to occur, there must be eight such particles. They are all massless spin-1 particles, and while they are electrically neutral, they do carry the color charge. Each gluon carries one color and one anticolor, although the color and anticolor need not necessarily be members of a corresponding pair. There is, for example, a gluon that carries the red and antigreen charge, another that carries blue and antired, and so forth.

The exchange of gluons between quarks holds the elementary particles together, just as the exchange of photons between electrons and protons holds an atom together. The only real difference is that the exchanges are a bit more complicated for the strong interactions because quarks must change their color charge when a gluon is emitted. Figure 25 shows one exchange process in a proton. The three quarks move along until the red quark emits the gluon that carries the red and antigreen (magenta) colors. This gluon then goes over the green quark, where the green and antigreen charges cancel each other, leaving the quark with a net red charge. The exchange of the gluon has thus interchanged the color charge of the two quarks but left the overall system with the correct combination to give white (color-neutral). You will note in this example that color charge, like electri-

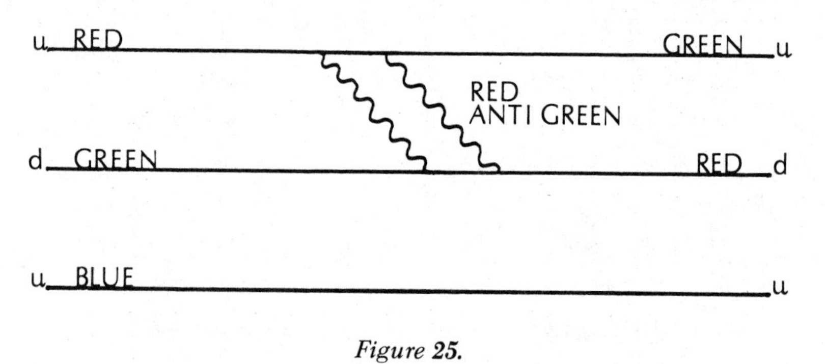

Figure 25.

cal charge, is conserved at each stage of the interaction. When the gluon is emitted, a red charge enters the interaction with the quark and leaves with the gluon, while the green of the final quark is canceled by the antigreen of the gluon. This is another example of the close analogy between color charge and electrical charge.

Quark Confinement

One of the major conceptual problems with the theory of strong interactions we have just outlined is the simple fact that no quarks have ever been seen in the laboratory. (There is one possible exception to this statement, which we will discuss in Chapter 12.) If every hadron is made up of quarks, it is difficult to see how this situation could arise; after all, we know atoms are made up of nuclei and electrons, and we see free nuclei and free electrons all the time. The only way that quarks could have evaded the intensive searches that have been made for them would be if, for some reason, they can exist only inside of particles and cannot exist as free entities. This property is called confinement, because the quarks are confined inside particles. It turns out that there is a very natural way to explain confinement in terms of gluon exchange. In the process of producing this explanation, we will also uncover a valuable way of describing the entire grand unification process.

The best way to approach the issue of confinement is to think of an electromagnetic analogy. If we have an ordinary electrical charge sitting in space, we know that the uncertainty principle allows a virtual electron-positron pair to be created in space nearby, as shown in Figure 26. The only requirement is that this pair recombine and annihilate in a time short enough to satisfy the principle. This sort of appearance and disappearance goes on all the time in the vacuum.

Near an ordinary charge, electrical forces will act on the virtual pair during its brief lifetime. If the original charge is negative, then the positron will tend to be pulled toward it and the electron will tend to be repelled from it. Since the creation and annihilation of virtual pairs is a continuous process, we can think of the original charge as being surrounded by a cloud of virtual electron-positron pairs, as shown in Figure 27. On the average,

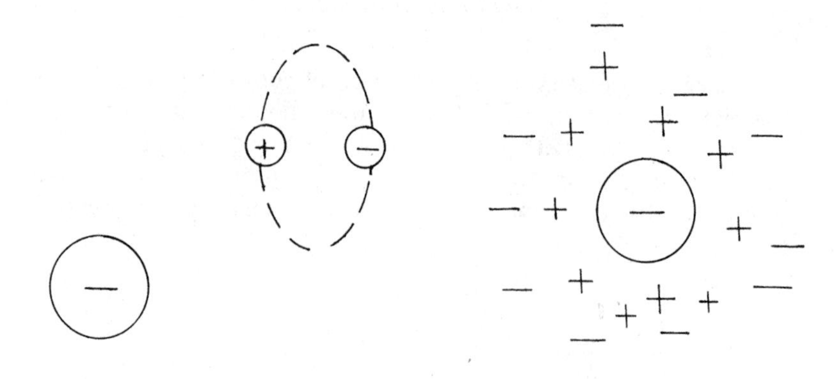

Figure 26. *Figure 27.*

this cloud of charged particles will arrange itself in such a way that the part of the cloud nearest the particle will be positively charged, while the part farthest away will be negative.

Now imagine another charged particle, such as an electron, coming near this particle-plus cloud system. If the electron stays far away, the system will look like an ordinary electrical charge. If, however, the electron has a high velocity, it will penetrate the system, forcing its way deep into the cloud. The higher the energy the electron has, the more deeply it will be able to penetrate. Therefore, if we are to describe the electron-particle interaction at high energy, we will have to think about what sort of forces act on the electron once it is in the cloud. There are two forces to consider—the one associated with the original negative charge and the one associated with the positive parts of the cloud near that charge. These two will tend to cancel out each other, with the positive cloud attracting the electron and the original charge repelling it. Thus, the total force on the electron will be less than that which would be exerted by the original particle alone. We say that the positive cloud shields the charge of the stationary particle. As far as the incoming particle is concerned, the situation is exactly what it would be if the original negative charge had a smaller value than it actually does.

The deeper we penetrate into the cloud, the more of the positive charge we leave behind us and the less of the positive charge we still see in front shielding the stationary charge. The total force on the electron exerted by the original charge and the cloud therefore increases. The higher the energy of the electron, the deeper it will penetrate; and the deeper it penetrates, the stronger the repulsive force it feels. It is as if someone were increasing the charge of the stationary particle as the electron gets closer to it. The effective strength of the electrical force, in other words, increases as the energy of the electron becomes greater. It must be emphasized that this effect has nothing to do with changes in any real charge but with the fact that the virtual cloud around a charged particle can shield some of that particle's charge.

Having made this point for electromagnetism, we can use the same technique in discussing the strong interactions. Any quark will be surrounded by a cloud of virtual quark-antiquark pairs. If the original quark is green, then we would expect a green-antigreen virtual pair to arrange itself so that the antigreen member of the pair is close to the original quark. Our first impulse, then, is to say that the shielding effects should be the same here as they are for electromagnetism and that the effective color charge of a quark should increase as the energy of the collision increases.

For the strong interactions, however, there is an important new factor in the argument. In addition to the quark-antiquark cloud, there will be a second overlapping cloud made of virtual gluons. An analogous cloud of virtual photons exists around an electrical charge, of course, but since the photons do not carry their own electrical charge, this cloud has little effect. In the case of the strong interaction, however, the virtual gluons in the cloud do carry color charge and hence have to be taken into account when we talk about the effective interaction strength. It turns out that the virtual gluon cloud produces exactly the opposite effect on the incoming particle as that associated with the quark-antiquark cloud. It provides a kind of "antiscreening," which adds to the color charge of the original quark rather than subtracting from it. You can think of this effect as being due to the green part of the gluon being "attracted" to the original green quark. The net effect of the two overlapping clouds, then, is shown in Figure 28. The original green quark is surrounded by a cloud of virtual particles, with the part of the cloud nearest the quark also having a green color.

Another quark approaching this system at high energy will penetrate the cloud. As it gets closer to the original quark (that is, as its energy increases), it sees less and less of the green part of the cloud. Consequently, it sees the effective color charge of the original quark becoming smaller and smaller. Thus, unlike electromagnetism, the strong interaction can be expected to appear

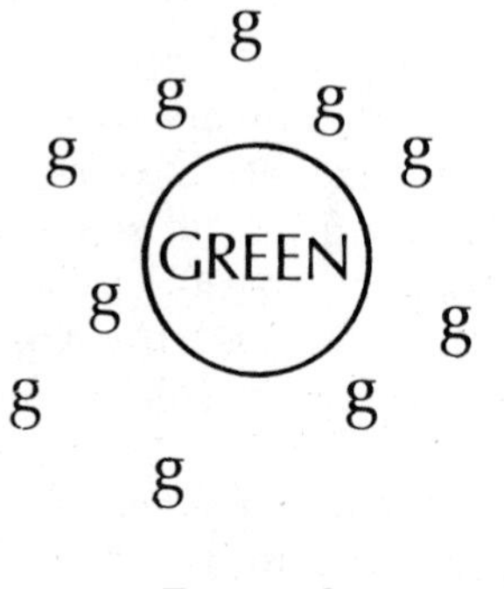

Figure 28.

less and less strong as the energy with which particles collide increases.

This result has two very important consequences. On the one hand, it means that if quarks are close together, as they are when they are bound into ordinary particles, they spend most of their time inside of each other's virtual clouds. Hence, they interact with each other as weakly as it is possible for them to interact. This property, known as the asymptotic freedom of quarks, means that we can picture the quarks in a proton as three marbles rattling around loosely in a container, with very little interaction.

On the other hand, if a quark should try to leave the proton, it will see the color charge of the system it is leaving behind getting bigger and bigger as it tries to fight its way out. You could imagine the quark as being tied to its original position by a loose rubber band. As long as it just rattles around inside the proton, nothing happens. But if it tries to get out, it finds itself pulled back more and more strongly the farther out it gets. In effect, then, the nature of the gluon cloud is such that no quark now bound into a particle can be pulled out. It is for this reason that we say the quark is confined in the particle and speak of the principle of quark confinement. This principle explains why no quarks are seen in the laboratory, since a quark cannot be freed from an elementary particle.

The Grand Unification

QCD, then, provides us with a theory of the quark interaction that satisfies the principle of gauge symmetry as applied to the color charge. The next question we ask is whether this interaction can now be unified with the electroweak force using techniques analogous to those presented in the previous chapter. It turns out that the answer to this question is yes, although the exact manner in which the unification must take place is still a subject of some debate among theoreticians.

The general scheme of things, however, is easy to describe. The interaction underlying all three forces (strong, electromagnetic, and weak) exhibits a gauge symmetry that allows us to change electrical and color charges at will at different points in space. Because we are now talking of combining the weak and strong interactions, we have to extend this symmetry and allow quarks to be converted to leptons and vice versa in exactly the same way. To compensate for these alterations, there must be a family of particles whose exchange is responsible for the unified force. Changes in these exchanged particles will exactly cancel out the changes in the original particles, leaving the resulting theory invariant. In the symmetric version of the theory, all of the exchanged particles would be massless. Because of the phenomenon of spontaneous symmetry-breaking, however, some of these exchanged particles will actually be massive, as the vector bosons were in the unification of the electromagnetic and weak forces.

The theoretical debate mentioned earlier has to do with the exact symmetry that this grand unified theory (GUT) obeys. This question is not yet settled, although the general results that any theory will produce are clear. For the sake of definiteness, I will outline here one of the simplest grand unified theories. (The reader who is so inclined may keep in mind that the actual theory that is finally developed may have a few more types of particles in it.)

The basic force operating in this grand unified theory is mediated by a family of twenty-four massless spin-1 particles. They play the same role in the grand unification that the four massless spin-1 particles did in the electroweak unification. Four of these particles, in fact, are to be identified with the four electroweak bosons, while eight more are to be identified with the gluons responsible for the strong force. The remaining twelve particles are denoted by the letter *X* and represent a new set of particles. The *X*-particles carry the color charge and an electrical charge of $\pm 1/3$ or $\pm 4/3$. Thus, there are the red, green, and blue *X* with charge $1/3$, the same colors with charge $-1/3$, and so on. In addition, these *X*-particles have the property that they can change a quark into a

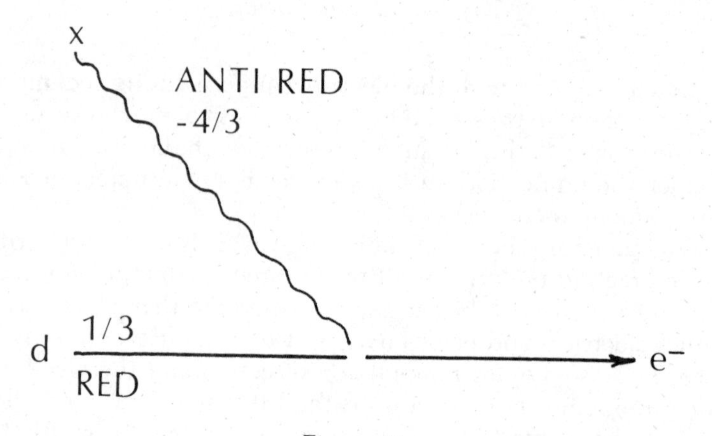

Figure 29.

lepton and vice versa. In other words, a process like the one shown in Figure 29, in which an *X*-particle combines with a quark to produce an electron, is possible in the grand unified theory, although it would not have been allowed according to traditional ideas of particle physics. This aspect of the grand unified theory is very important in terms of the predictions the theory makes about quantities that can be experimentally tested.

In Chapter 6 we saw that the process of spontaneous symmetry-breaking converted some of the energy of the system into mass, just as the analogous process in iron converts some energy into large-scale magnetic fields. In the case of the twelve *X*-particles, they acquire a huge mass—10^{15} GeV. This is a quadrillion times the mass of the proton, and roughly the mass of a human blood cell. The idea that there might be particles that are macroscopic in mass but microscopic in size has caused considerable rethinking of old ideas in physics, as we shall see later.

When the energy involved in collisions between particles is higher than 10^{15} GeV, then, the fact that the *X* is massive and the gluons are not becomes irrelevant and the underlying symmetry of the system is made manifest in nature. When the energy drops below this point, the effects of spontaneous symmetry-breaking will be seen and the strong force will appear to be very different in character from the electroweak. For reference, the mass of 10^{15} GeV is called the *grand unification mass,* and the corresponding energy, the *grand unification energy.*

Energies below the grand unification energy are still very large compared to the masses of the vector bosons, of course, so even after the energy has dropped below 10^{15} GeV, no effects of the differences in mass between the photon and the vector bosons will be apparent. The symmetry associated with the electroweak force will be manifest in nature until the energy of collisions drops below 100 GeV or so, at which point the three forces involved in the grand unification will all appear distinct.

There is another way to think about the grand unification. We saw earlier that as the energy of interaction increased, the effective charge of particles and quarks changed because of the presence of virtual clouds. We can represent these changes on a graph as shown in Figure 30. The electroweak coupling constant most closely analogous to electrical charge increases slowly as the energy of collisions increases, while the strength of the strong interaction drops. The two coupling constants representing the strengths of the two different interactions approach each other slowly, becoming equal at the unification energy. We can think of the unification, then, as the point at which the forces become identical in their strengths, as well as the point at which the particles being exchanged to generate the force all have masses small compared to the collision energy.

The most striking thing about this graph (and about the grand unification in general) is the enormous disparity between the energies at which the two unifications occur. For example, in Figure 30 we have had to compress the energy scale to make everything fit on the page. If we had actually drawn the entire thing to scale, the point at which the two curves would join would be somewhere out beyond the orbit of Mars. The theory seems to tell us that once we get beyond 100 GeV or so, we can expect nothing interesting to happen in particle physics until we

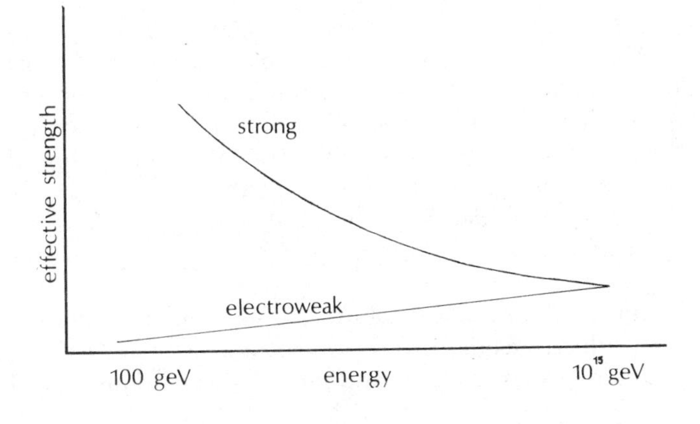

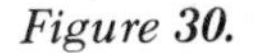
Figure 30.

get to the unification energy. Physicists call this vast, rather dull energy region the great asymptotic desert. Why the two unifications should occur at such totally different energies is a fundamental problem in theoretical physics, and at the moment no one has any good ideas on how to solve it.

Successive Freezings

We have repeatedly used the analogy of the formation of a magnet or a snowflake to talk about symmetries in nature. The analogy also gives us a very nice way of visualizing the unifications of forces. When a freezing occurs, we ordinarily think of a rearrangement of the matter that makes up a system. A snowflake, for example, is made when water molecules lock themselves into rigid crystals, and a magnet is made when iron atoms lock themselves into alignment with each other. No such dramatic event occurs when a pair of forces become unified. The matter in the system stays pretty much as it was; it is only the interaction between bits of that matter that changes.

From the point of view of symmetry, however, the description of freezing and the description of unification are much closer. Liquid water is the same no matter how we rotate the water or ourselves. Frozen water, on the other hand, is only invariant if we rotate through some multiple of 60°. Therefore, the process of freezing can be thought of as one in which the symmetry of the system is reduced. From a system in which there is no preferred direction (that is, a system of high symmetry) at 33°F, water goes to a system in which there are very definite preferred directions (a system of lower symmetry) at 31°F.

The analogous process in the unification of forces is easy to see. Above the unification energy, forces are unified. This is a situation of high symmetry as far as the forces are concerned. Below the unification energy there is a distinction between forces, and the system is one of lower symmetry. We can therefore think of the process as the "freezing out" of the two separate forces from the original unified force. The only thing to keep in mind is that this is an abstract sort of freezing and does not involve a rearrangement of matter.

In terms of this visualization, we can think of the strong, weak, and electromagnetic forces as being analogous to three different fluids—water, alcohol, and mercury, for example. At high temperatures, all three materials will be fluids, and we would say that the system has a high degree of symmetry. As we lower the temperature, however, this symmetry does not survive. At 32°F, the water will freeze, leaving the mercury and alcohol as liquids. This corresponds to the "freezing out" of the strong force at 10^{15} GeV. Below 32°F we have a much lower symmetry in our system: two of the materials are still liquid, but the third is a crystal and would appear to be completely different from its fellows. At −40°F the mercury would freeze. This would correspond to the "freezing out" of the weak force at 100 GeV. Below this point the symmetry of the original system is completely lost. Two of the materials are now solids of quite different structure while the third remains a liquid. An observer looking at this final configuration would be hard put to realize the basic similarity of the three materials, and yet that basic similarity would be manifest at temperatures above 32°F. In just the same way, we now look at the strong, weak, and electromagnetic forces and perceive them as being entirely different from each other, and yet at very high energies our theories tell us that they must be identical.

The Ultimate Test of the Grand Unification Theory

The energies involved in the grand unification theory are so high that there is no hope that we will be able to reproduce them in our laboratories at any time in the foreseeable future. Consequently, we have to rely on indirect tests to find experimental verification of the theory.

The most striking feature of the GUT is the prediction that at very high energy it is possible to change quarks into leptons and vice versa through the mediation of the *X*-particles. This means that particles made up of quarks, like protons and neutrons, ought to be unstable. It should be possible to turn one of the quarks into a lepton (electron, positron, or neutrino), causing the proton to decay spontaneously. Because the mass of the *X*-particle is so high, we expect this process to be very unlikely, but the theory nevertheless predicts that it ought to occur. Since no other theory makes such a striking prediction, measuring the stability of the proton turns out to be an important test of the idea of grand unification. The search for the decay of the proton is such a crucial undertaking at this stage of our understanding that we will devote all of Chapter 8 to a detailed description of the theoretical and experimental effort that has been expended. For the moment, we note that although the grand unification mass is indeed beyond our reach, we can still test the grand unification idea in our laboratories here and now.

Summary

*The process of unification in the fundamental forces can be taken one step farther. By the mid-1970s, a theory of the strong interactions had been developed. It is called quantum chromodynamics (*quantum *refers to the fact that we are dealing with particles,* chromo- *to the fact that the fundamental strong interaction depends on the color charge of the quarks). In this theory, the strong force is mediated by the exchange of massless particles called gluons between the quarks. Unlike the photons, which mediate the interaction between electrically charged particles but do not themselves carry an electrical charge, the gluons carry the color charge. Because of this fact, some simple arguments lead us to the conclusion that the force between two quarks will become stronger and stronger as the distance between the quarks increases. Thus, a quark trying to leave an elementary particle will find itself having to overcome an ever-increasing resistance. The consequence of this fact is that quarks are "confined" within the particles and cannot be seen in isolation. This explains the failure to see them in the laboratory.*

The unification of the electroweak and the strong forces takes place in the following way: At very high energies, when the underlying symmetry of the system is manifest in nature, the strong-electroweak force is mediated by the exchange of twenty-four particles. In the perfectly symmetrical case, all of these particles are massless. Spontaneous symmetry breaking, however, causes twelve of these particles (called the X*-particles) to have a mass around* 10^{15} *GeV. This is called the grand unification mass or the grand unification energy. At energies much higher than this, we can ignore the mass of the* X *and say that all twenty-four particles being exchanged have essentially zero mass. In this case, we would see the forces as being identical. At energies below* 10^{15} *GeV, we see the effects of spontaneous symmetry-breaking and perceive the strong force to be different from the electroweak.*

The other twelve massless particles are the eight gluons, which mediate the interaction between quarks; the three vector bosons, which mediate the weak interaction; and the photon, which mediates the electromagnetic interaction. We say the forces are unified, because in this picture all three forces are seen to arise through the exchange of the same family of particles. The process by which the forces become separated as the energy of the interaction falls is analogous to a series of freezings, and the reader is urged to read the section entitled "Successive Freezings" on page 122 to get an understanding of this imagery.

The X*-particles have the property that when they are absorbed by a quark, they convert it to a lepton and vice versa. They give rise, therefore, to the most important prediction of the grand unified theories—that protons are unstable. Chapter 8 will deal with this concept in detail.*

Chapter
8

The Decay of the Proton

They also serve who only stand and wait.
JOHN MILTON
"On His Blindness"

The most astonishing aspect of the grand unification theory (GUT) may not be its intellectual beauty and simplicity but the predictions it makes about the long-term stability of matter. And more amazing than these theoretical predictions is the fact that it is possible, with experiments being done as you read this, to test whether the predictions are true or not. These experiments are important not only because they provide us with a way of testing the whole concept of grand unification but because they give us a way of learning about the ultimate fate of the universe. For the simple fact is that the theories we have been discussing make an unambiguous prediction: All matter, even those forms that seem permanent to us, is inherently unstable and will eventually decay.

The Stability of Matter

From our knowledge of elementary particles, we know that there are only a few that might qualify for the adjective *stable.* By far the greatest number of known particles decay into other particles on time scales short enough for us to detect and measure the disintegration process. From the long-lived isotopes of uranium to the most evanescent hadron, the rule that seems to operate is that if it is possible for a particle or nucleus to decay (that is, if the decay is consistent with other laws of nature), then it will do so.

There are few laws that would put constraints on decay. The most important are the conservation of energy and the conservation of electrical charge. The first of these simply says that the energy of mass and motion of the decay products must be equal to the energy locked up in the mass of the original particle. To take one example, this implies that a particle's decay products cannot be more massive than the particle itself. The second law operates in much the same way: it tells us that the electrical charge must be the same before and after a given decay, and excludes, for example, the decay of a positively charged particle into a set of particles whose net electrical charge is zero.

These two requirements guarantee that the electron cannot decay, for there is no particle or set of particles that is both less massive than the electron and that bears a negative electrical charge. They do not, however, imply that the same must be true for the proton, for there are many combinations of particles less massive and positively charged. If we had only these two restraints, it would be theoretically possible for the reaction

$$p \rightarrow e^{+} + \pi^{\circ}$$

to occur. That no theory up to now has ever considered such a reaction is not because it is logically impossible but rather because protons seems to be as stable as the electron. This has led physicists to add the law of baryon conservation to the laws of conservation of energy and charge, but that law has always been in the nature of something added "by hand." While there are many phenomena in nature that cannot be explained without the conservation of energy and electrical charge, the main phenomenon "explained" by strict baryon conservation is the apparent stability of the proton. If the number of baryons in an interaction has to be conserved and if the proton is the lightest baryon, then it follows that the proton cannot decay.

One measure of the stability of the proton is a quantity called its lifetime. This is the time it takes for about ⅔ of a given sample of protons to disintegrate. From the fact that the universe is still in existence 15 billion years or so after the Big Bang, we know that the proton lifetime cannot be a great deal less than 10^{10} years. This is a long lifetime, but not extraordinarily so. Uranium 238 has a half-life of about 5 billion years. But as Maurice Goldhaber of Brookhaven National Laboratories has pointed out, we "know in our bones" that the proton lifetime must be considerably longer than the age of the universe. The calculation to confirm this is simple. The average human being has roughly 10^{28} protons in his body. If the proton lifetime were only 10^{10} years, then your protons would be disintegrating at the rate of 30 billion per second. In this case, each of us would be a radioactive source of strength 1 curie, giving off as much radiation as a handful of material from the core of a nuclear reactor. Pursuing this line of reasoning, if we require that the body be less radioactive than medical tracers, we can conclude that the lifetime of the proton must be at least 10^{16} years—a million times as long as the age of the universe. For reference, if the proton lifetime were indeed 10^{16} years, and you had been alive since the Big Bang, you would have lost about a thousanth of an ounce in body weight because of the decay of your body's protons.

This sort of simple argument shows why the proton is usually considered to be stable; we know that it must live a very long time indeed. It also shows that the normal methods of looking at particle decay, most of which involve tracking an individual particle until it disintegrates, just will not work for the proton. We would have to wait too long. The alternative is to watch a lot of protons and hope we will catch one of them in the process of decay. For example, if a single proton takes 10^{16} years to decay (on the average), in a collection of 10^{16} protons, one will decay each year (on the average).

This sort of argument about the proton lifetime also illustrates another important point. In any determination of this quantity, there must be some method of detecting the products of the decay. In our discussion so far, we have used the human body itself as the "detector," pointing out that if the radiation levels were as high as they would have to be for a proton lifetime of 10^{16} years, we would all be dead from radiation-induced cancer. The fact that you are reading this book is taken as evidence that your detector has not absorbed radiation at that level.

There are, of course, much more sensitive detectors of radiation than the body. If it were possible to place instruments in the body to give an instantaneous readout of radioactivity, we could get a much more accurate (and higher) limit on the proton lifetime. Although we cannot do this for the body, we can certainly do it for an inanimate object like a quart of water. Since the logic of what follows is typical of all proton decay experiments, we will go through it in some detail.

A single proton or neutron has a mass of 1.67×10^{-27} kg. A liter of water (about a quart) has a mass of 1 kg, and since the mass of any material object is made up almost entirely of the mass of the protons and neutrons it contains, we conclude that there must be $1/(1.67 \times 10^{-27}) = 6 \times 10^{26}$ protons and neutrons in a liter of water.* If the proton lifetime is 10^{16} years, then we would expect protons in the water to decay at the rate of 6×10^{10} decays per year, or roughly 2,000 per second. If we put a bunch of Geiger counters into the water, we would expect them to start clicking rapidly, indicating a high level of radioactivity in the sample. In fact, a Geiger counter placed in or near a quart of water will not register a high level of activity but will click about once a second or so.

Even this level of activity is due to cosmic rays, but let us ignore that for the sake of argument. Suppose at that point we just went ahead and assumed that we were measuring one proton decay per second in our sample. What sort of proton lifetime would that give us? One decay per second corresponds to about 3×10^7 decays per year. In a sample of 6×10^{26} particles, this corresponds to a lifetime of $6 \times 10^{26}/3 \times 10^7 = 2 \times 10^{19}$ years. This is already a factor of 500 better than the limit we set by using the human body as a detector.

But even this limit is not accurate enough, because, as we intimated, virtually all of the counts that we attributed to proton decays were, in fact, due to the influence of cosmic rays. In practical terms, we would discover this fact by noting that our Geiger counter was going off once a second or so, whether it was in the sample of water or not. This would lead us to suspect that the clicks were not actually the record of decays and we could verify this hypothesis by putting the counter under a stack of lead bricks, which would screen out cosmic rays. The number of counts would go down dramatically, whether the counter was in the water or not. From this we conclude that if we intend to

*Perhaps it would be a good idea to clear up one possible source of difficulty at this point. We have spoken loosely of "proton decay," but as we shall see shortly, both protons and neutrons bound into nuclei can undergo the decay process predicted by the GUT. This fact may prevent some confusion over factors of 2 in the above and subsequent calculations.

make a serious attempt to detect radioactivity associated with proton decay, we had better find some way of keeping cosmic rays out of the experimental area. The only way that this can be done is to have enough material over the apparatus to absorb cosmic rays (and particles produced in their collisions), keeping them away from the instruments. Given the penetrating power of cosmic rays, the only practical way to provide adequate shielding is to put the entire experiment in a mine.

In fact, the best limit on the proton lifetime prior to the current spate of interest in the subject was obtained as a secondary result of an experiment designed to record cosmic ray and neutrino interactions deep underground. Carried out by groups from Case Western Reserve University, the University of California at Irvine, and the University of Witwatersrand, the apparatus consisted of 20 tons of material in a gold mine more than 2 miles underground in South Africa. After making an analysis of their results that is similar in spirit to (but much more complicated than) our analysis of the quart of water, the experimenters concluded that the proton lifetime had to be greater than 10^{30} years. I say "greater than" because they saw no events that could be ascribed to proton decay, and the sensitivity of their instruments would be such that they would have seen such events if in the proton lifetime was less than 10^{30} years.

This is a time so long as to be virtually unimaginable. The age of the universe, remember, is "only" 10^{10} years or so. If you had been alive since the Big Bang, only 100 million of the protons in your body would have decayed by now, a mass much less than that of a single cell and approximately that of a few viruses or a hundred run-of-the-mill protein molecules. In terms of experimental design, a lifetime this long means that if we want to see one proton decay per year we have to instrument and monitor a sample of 10^{30} protons. This is roughly $1.67 \times 10^3 = 1.67$ metric tons of mass, about the equivalent of a small swimming pool full of water. If we want to improve on this measurement, we have to watch a proportionately larger number of protons and, therefore, a proportionately larger amount of material.

This was the experimental situation in 1974 when Steven Weinberg, Howard Georgi, and Helen Quinn (then all at Harvard University) proposed the grand unification scheme and made the first quantitative prediction for the lifetime of the proton. Their prediction was that the proton should decay with a lifetime of 10^{32} years. A later refinement reduced this number to about 10^{31}

years. The exact number is not nearly so important, however, as the astounding coincidence that these calculations imply. Just as theorists produced a theory that is capable of predicting the rate at which a proton can be expected to decay, experimentalists completed work that showed that the limits placed on this process are only slightly less than the prediction.

What this means is that experimentalists knew that if they could design an apparatus that would be sensitive to proton decay at the level of 10^{31} to 10^{32} years, they would be assured of getting an interesting result. Either they would measure a fundamental new process, or they would prove the theoretical calculations wrong. (As a theoretical physicist, I have never been clear in my own mind as to which of these prospects presents the greatest attraction to my experimental colleagues, but that is beside the point.) The combination of the technological challenge and the theoretical predictions simply proved irresistible, and the race to detect the decay of the proton was under way. Before we look at some of the ingenious schemes that have been proposed to carry out this task, it would probably be a good idea to make a small digression and discuss exactly how it is that the GUT predicts the decay of the proton.

The Theoretical Picture

The picture of the proton in the GUT is actually quite simple. There are three quarks bound together by an interaction that involves the exchange of gluons. In terms of the type of diagrams we have used to discuss the fundamental forces, the proton would have the appearance of Figure 31. Two *u*-quarks and a *d*-quark move along, exchanging a gluon once in a while. So long as the proton is stable, this is all there is to it.

In Chapter 7, however, we learned that there are other kinds

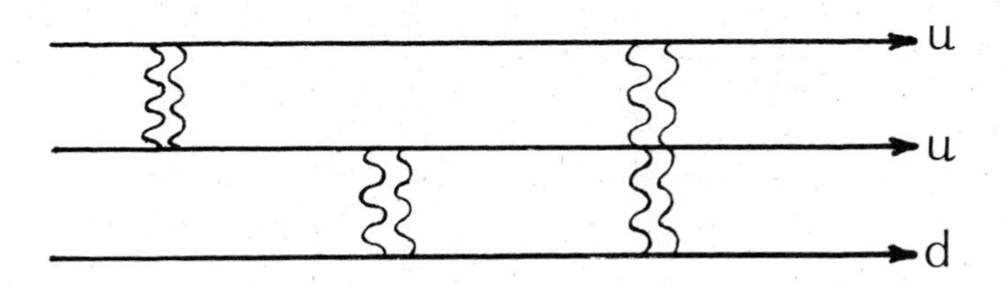

Figure 31.

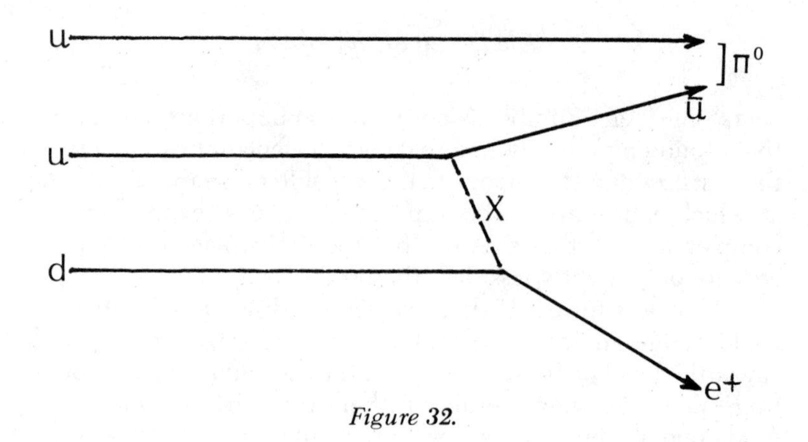

Figure 32.

of particles that can be exchanged between quarks. The *X*-boson is one such. From the properties of the *X*, we know that a process like the one depicted in Figure 32 is theoretically possible: one of the *u*-quarks emits an *X* of charge $\frac{4}{3}$, turning into an anti-*u*-quark (charge $-\frac{2}{3}$) in the process. This process violates baryon conservation, of course, but is consistent with the conservation of both electrical charge and energy. The anti-*u*-quark combines with the other *u*-quark to form an ordinary neutral meson, while the *X* interacts with the *d*-quark to produce a positron. Again, baryon conservation is violated at this interaction, but nothing else. The net result of this interaction is that the original proton is converted into a neutral meson (such as the π°) and a positron.

Alternatively, we could have a situation like the one in Figure 33: This would correspond to the decay of the proton into a

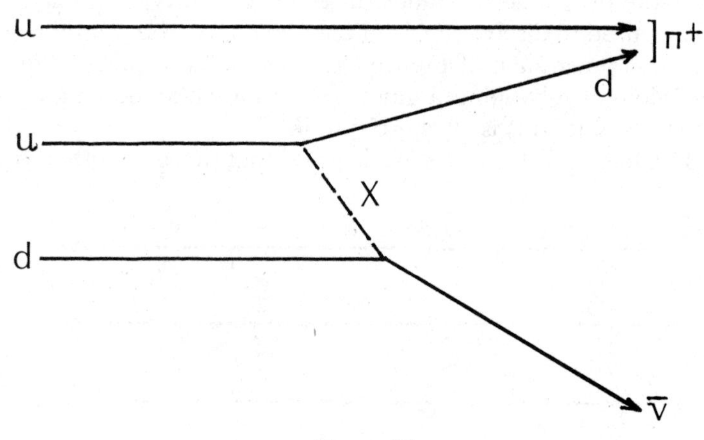

Figure 33.

positively charged π meson and an antineutrino. In this case, it is the X with charge $+\frac{1}{3}$ that is exchanged.

The reason that the proton should be almost stable is clear from these diagrams. Any process leading to proton decay must necessarily involve the exchange of an X-particle, and the X-particle has a mass characteristic of the grand unification—10^{15} GeV. Just as the weakness of the weak interactions is directly attributable to the large mass of the vector bosons, so too is the extreme slowness of proton decay attributable to the high mass of the X-particle.

These diagrams (and others like them) also suggest just what a proton decay would look like in the laboratory (see Figure 34). A proton in an isolated sample will decay, producing, for example, a π° and a positron. Ordinary detectors will pick up the latter, while the former will travel a short way before it decays into two photons, which can also be detected with appropriate instruments. Thus, in this case, we would recognize a proton decay by seeing a group of particles with net charge $+1$ appear in our apparatus without anything entering from the outside.

The last phrase is very important. If an uncharged particle (a neutrino, for example, or a neutron produced by a cosmic ray collision near the experiment) entered the apparatus, we could get a situation like that in Figure 35. The entering neutrino, being uncharged, would not be seen, even though we have indicated its presence by a dotted line. When the neutrino hits a proton, it produces a positron and a resonance, which quickly decays into a π° and a neutron. Some of the time the neutron will be moving slowly, so that unless some care is taken in the design of the

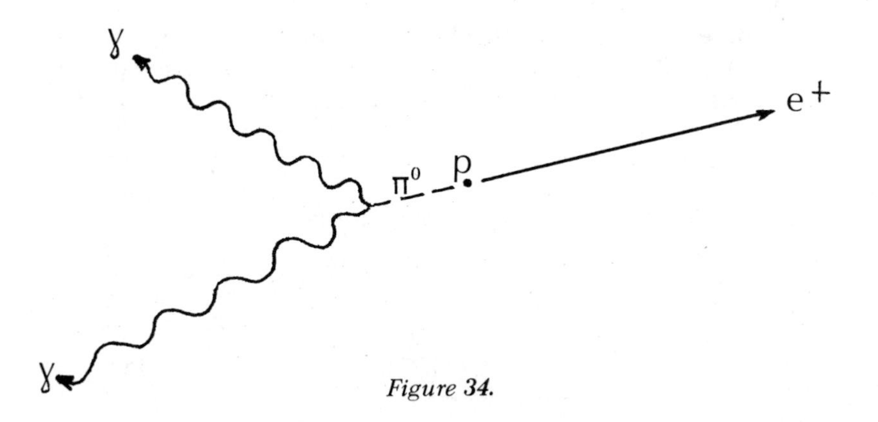

Figure 34.

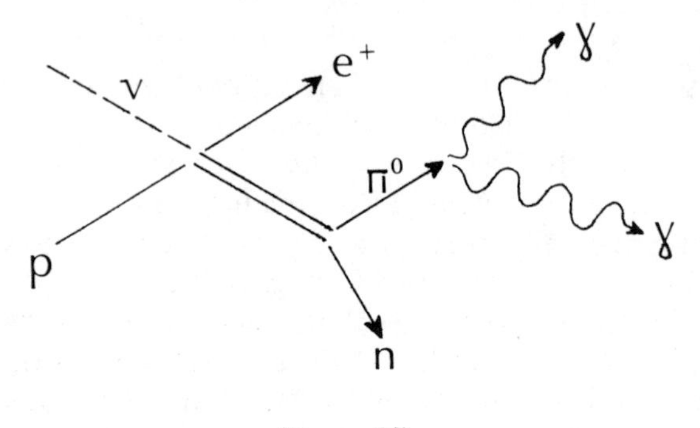

Figure 35.

apparatus, it will not be seen. If this happens, then the experimenter will see exactly the same thing in the neutrino-induced event as he did in the genuine proton decay. This illustrates the complexity of the problem.

A Survey of the Experiments

In the following table, we show the amount of material that must be monitored if we are to see one proton decay per year. In each case we give also the length of the side of the cube of water that would be required to hold that many protons. The table shows that one can quickly get involved with prodigious amounts of material in the proton decay game. For example, the amount of water needed to see one decay per year at a lifetime of 10^{32} years is equivalent in volume to a small house. The numbers in the table will help us to understand the sheer scale of the experiments now running.

There are two general schemes for detecting the decay of the proton, differing in their choice of material to be monitored. One class, the so-called dense detectors, use some heavy material,

Assumed lifetime (years)	Mass of Material (metric tons)	Side of Cube of Water (feet)
10^{30}	1.67	3.75
10^{31}	16.7	8.25
10^{32}	167.0	18.0

like steel or concrete, as the source of protons. This has the advantage of compactness; a 30-ton detector could fit into a small room. In addition, it is easier to make the decay products stop inside the detector volume, an important technical feature. On the other hand, the density of the material means that instruments for detecting decay products must be fairly closely spaced throughout the apparatus. Water detectors require a larger volume to get the same number of protons. Very pure water, however, is remarkably transparent. Consequently, they can use something called Čerenkov radiation to detect particles over long distances and need not have the dense array of instruments characteristic of dense detectors.

When a charged particle travels through any medium at a speed exceeding the speed of light in the medium (but *not* the speed of light in a vacuum), it emits light in a cone as shown in Figure 36. This emission is analogous to the sonic boom emitted by aircraft traveling faster than the speed of sound. Light emitted in this way is called Čerenkov radiation, after the Russian physicist Pavel Čerenkov, who was awarded the Nobel Prize for its discovery in 1958 (the discovery itself was made during the 1930s). If a proton decay in a tank of water results in the emission of fast particles, then each such particle will emit Čerenkov radiation. If the water is pure enough, it will be possible for detectors to record this radiation over great distances. Thus, the advantage of the water detector is that instruments for recording decays need not be densely packed in the water but can be spread out. In some versions of the experiment, the detectors may be put along the walls of the water tank, leaving the central mass free of obstacles.

In Figures 37 and 38, we show a highly simplified sketch of

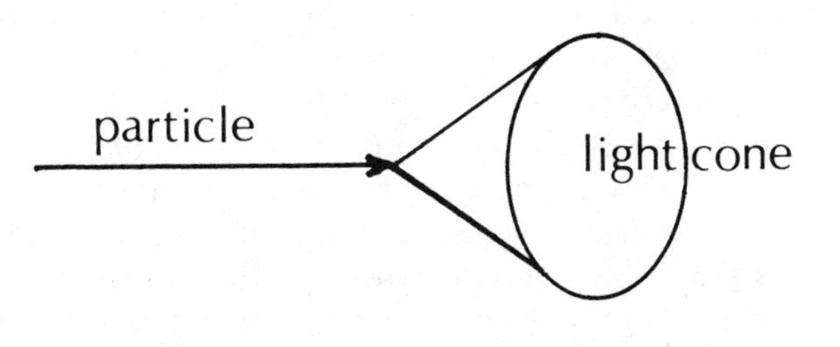

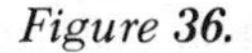
Figure 36.

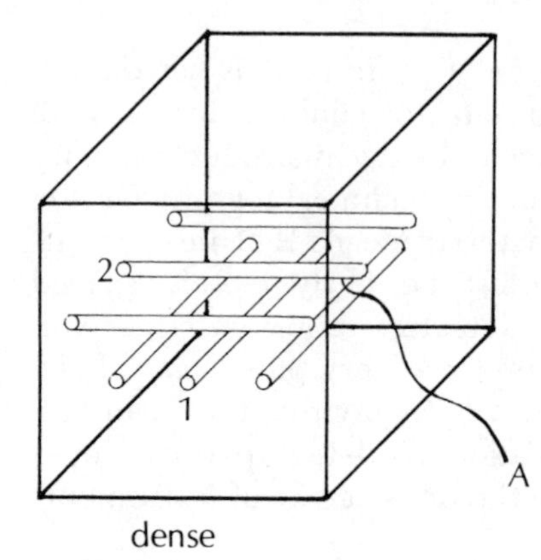

Figure 37.

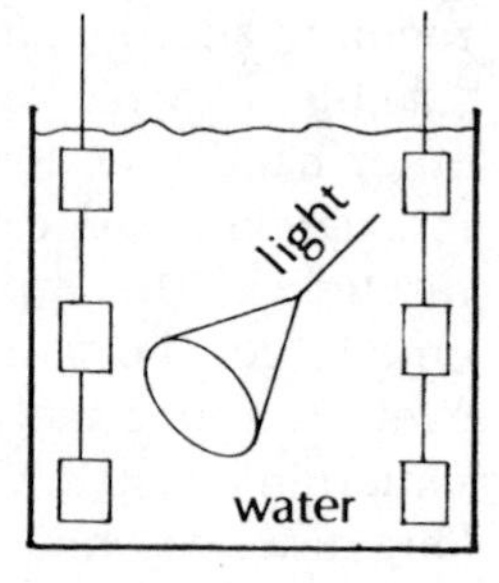

Figure 38.

how each of these major types of detectors would look in operation. Figure 37 is a dense detector. It consists of slabs of concrete thickly interspersed with particle detectors. These detectors are long tubes of gas, something like a fluorescent light bulb in appearance. When a particle passes through the tube, an electrical signal is generated. The tubes are arranged so that each layer is at right angles to the one above it. A particle moving through the layers will trigger tubes in succession, thereby leaving in its wake a record of its progress. The alternation of directions of the tubes makes it possible to trace out its path. For example, if the tube labeled 1 fired first, followed by the tube labeled 2, we would know the particle had been in the area defined by the intersection of the two tubes; the point labeled A.

The water detector operates slightly differently, as shown in Figure 38. Here a particle moving through the apparatus emits a cone of light that is detected by several phototubes. From the location of these phototubes and the angle of the light reaching them, the location of the particle can be deduced.

At first glance, it would appear that setting up a proton decay experiment is easy, at least in principle. You just have to find a suitable underground location, assemble the appropriate number of protons, instrument the system, and wait for something to happen. As you might expect, however, the reality of the situation is considerably more complicated than the theory. We have already seen that the possibility that uncharged particles can enter the detector introduces the rather difficult problem of separating out events that look like proton decays but are not. This problem is especially acute near the edges of the detector, since either cosmic ray collisions or ordinary radioactive decay in the neighboring material can generate particles that can produce signals similar to those detected in a proton decay. This means that any event near the edges of the mass being monitored has to be discarded, and only those events in which the particle tracks lie wholly within the central mass (the so-called fiducial volume) of the detector are taken to be reliable.

For the case of water detectors, this means that events within six feet or so of the walls of the water tank have to be discarded. Referring to our table, we see that the largest volume of water shown, a cube 18 feet on a side, would have a fiducial volume equivalent to a cube only six feet on a side, or only 4×10^{30} protons. The "edge effect," therefore, requires that the volume of material in a real experiment be much larger than those shown in the table. Indeed, the largest water detector in existence contains some 10,000 tons of water, corresponding to a tank as big as a five-story building. This corresponds to more than 10^{33} protons in the fiducial volume, enough to produce 100 decays per year (or about 2 per week) if the lifetime is 10^{31} years.

For dense detectors, an alternate approach to the edge problem is possible. The detector, because of its relatively small size, can be surrounded by counters that will record the passage of stray particles. In this way, events created by such particles can be identified and thrown out (or *vetoed,* to use the jargon of experimental physics).

The edge effect explains why the earliest claim for the detection of a proton decay was greeted with much skepticism, if not downright disbelief, by the scientific community. This experiment was a collaboration between Indian and Japanese physicists, using a dense detector consisting of 150 tons of iron located 12,000 feet underground in the Kolar Gold Field in India. They saw a few events of the type in Figure 39—two or more tracks emanating from a point, with at least one track touching the edge of the detector. There are two possible interpretations of this sort of event: one is that a genuine proton decay occurred at point *A*; the other is that a particle entered the apparatus from the left and collided with a nucleus at point *A,* either being deflected itself or producing another charged particle in the process. Had there been a counter at the side, it would have been possible to rule out one or the other of these alternatives, leaving us with an

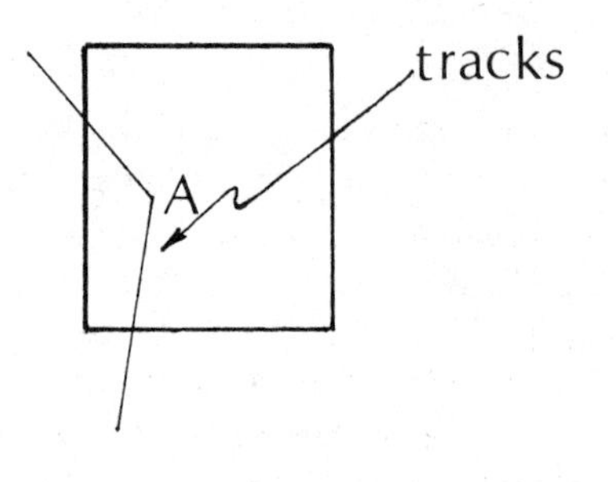

Figure 39.

unambiguous interpretation of the event. As it is, the events of this type cannot provide solid evidence for proton decay.

DOWN IN THE MINE

The requirement that a proton decay experiment be shielded from cosmic rays says that the experiment has to be done in a mine or deep tunnel—someplace under a few thousand feet of rock but accessible enough so that the equipment can be installed and then kept running over a period of several years. Telling an experimenter to "find a mine" is reminiscent of the old Kentucky recipe for bear stew that begins, "First, get a bear." It is easier said than done. In Europe there are a few tunnels in the Alps (most notably those under Mont Blanc) that have enough of an overburden of rock to serve as a suitable cosmic ray shield, and European physicists are fortunate enough to be able to set up shop in old caverns that were blasted out for use as equipment storage depots in these tunnels. In North America, however, we have no such luck. When the interest in proton decay experiments heated up in the late 1970s, there was a furious nationwide scramble to find places to work, a process that necessarily involved a search for space in a deep mine.

There is a thriving mining industry in America, of course, but much of it involves surface minerals that can be recovered by opening shallow pits or strips. Such mines, open to the air, are obviously not suitable for the proton decay experiment. Coal mines, although common, present a different problem, in that the ever-present danger of explosion rules out the use of the kind of electronic equipment the experimenters would have to install. This leaves two possibilities: salt mines, which usually have large caverns hollowed out by the mining process, and metal mines, the value of whose metal justifies the high cost of burrowing deep into the earth. Both types are being used as homes for proton decay experiments at the time.

Salt is one of the cheapest materials you can buy; a pound can be had for about 25 cents in your supermarket. This means that salt has to be mined fairly close to the place where it will be processed and used, otherwise the transportation costs would price it out of the market. There is a mine, operated by Morton Salt, a division of Morton-Thiokol Company, located near the shores of Lake Erie in Ohio. The mine itself actually extends out under the lake. When physicists from the University of Michigan, the University of California at Irvine, and Brookhaven National Laboratories first began to investigate this site as a possible place to build the world's largest water detector, it seemed ideal in many respect—close to major transportation centers, 2,000 feet underground, and run by a company willing to cooperate with their project. Unfortunately, there was no chamber in the mine big enough to house their detector.

Then, in one of those chance events that comes along just often enough to keep us believing in happy endings, the Dosco Company, a manufacturer of heavy mining equipment, arrived on the scene with a new machine that they wanted to demonstrate to Morton. The machine was put to work drilling out a cavity that could contain a five-story building, which it did with dispatch. As a side product, the physicists acquired $60,000 worth of raw salt that they sold back to Morton. They then set to work installing a huge plastic liner to hold their water. In August 1982, after some trouble with leaks, the liner was filled with 10,000 tons of purified water, the 2,048-Čerenkov detectors were turned on, and the long wait for the proton decay signal began.

THE ALL-MINNESOTA DETECTOR

The problems that faced Marvin Marshak, head of a proton decay experiment centered at the University of Minnesota, were somewhat different. The Minnesota group was the first to have a proton decay experiment up and running in North America. They had originally been involved in the search for a mine when they were part of a larger group. The story of the negotiations that took place in the latter part of 1978 and through 1979 give us an insight into the way that big-time physics is carried out these days. Teams from the various universities involved in the conceptualization and design of the experiment would fly into Chicago, meet for a day, and return to their home bases on the evening flights, for all the world like a bunch of sales executives. As the negotiations progressed, there were lawyers and contracts enough to satisfy the most cautious legal adviser. I have to admit that I was a little surprised that "ivory tower" physicists would operate in this way, but then I remembered that large proton decay experiments such as the one under Lake Erie cost over $3 million to build and operate—enough money to make all the advance negotiations worthwhile. (I am afraid that the image of the lonely scientist working at his bench is a thing of the past, at least in large-scale particle-physics experiments.)

In any case, the group began a search for suitable mine space at about the same time. As Marshak explains it, "We were faced with a real dilemma. When a mine is abandoned, it deteriorates quickly, and the cost of renovating and maintaining an abandoned mine was far out of our reach. In working mines, on the other hand, miners are often paid according to some incentive system, so that having a bunch of physicists interfere with their work would cost them money. This isn't a situation we'd like to be in."

Their search took them to the offices of federal and state mine inspectors, to professors of mining, and to anyone who might know of a suitable place for their experiment. In each place they asked, "Do you know of a mine we could use?," and in each case the answer was no. Then, in 1978, about the time that the Minnesota group struck out on its own, Marshak remembered a place he had visited in northern Minnesota, near the town of Soudan, roughly two hundred miles north of Minneapolis. There was a state park there to commemorate one of the first iron mines in Minnesota, a mine that had stopped commercial operations in 1963. The mine was maintained by the state of Minnesota, and tour groups were taken through during the summer. It was, in other words, a mine that was maintained but not worked—the perfect combination for the experiment Marshak had in mind. The university quickly arranged to use one of the chambers in the mine, a little over two thousand feet below ground, for this purpose.

While the search was in progress, the Minnesota group had decided that it was more likely to achieve its goals with a dense detector system than water, so it started work on a small (thirty-ton) prototype detector. It convinced a local company to donate a couple of truckloads of iron ore. Hauling the ore back to the university, the scientists bought a Sears Roebuck cement mixer, hired a few students, and set up shop in an old building next to the Mississippi River. Mixing the ore with concrete, they poured 432 slabs, each 10 feet by 1 foot by 1.5 inches (the size of the slab being dictated by what two students could handle conveniently). Each slab had eight steel tubes in it, about the size of a vacuum cleaner hose. These tubes would hold the counters needed to detect the products of proton decay. The entire apparatus was trucked to Soudan, taken down into the mine, assembled, and tested. By the fall of 1981, the detector started taking data.

PROSPECTS

In January 1983, the Irvine-Michigan-Brookhaven experiment located in the salt mine under Lake Erie announced the results of their first run. In eighty days, no candidate events for the decay of a proton were seen in their apparatus. From this fact they concluded that the lifetime of the proton must be longer than 6×10^{31} years. More recent results, based on still more running time, have pushed this limit to near 10^{32} years, with no decays seen. Lawrence Sulak of the University of Michigan commented that, based on this result, it would take the group about three years to push this limit to 6×10^{32} years. If no decays were seen during this period, it would be unlikely that proton decay would be detected in the near future. The reason is that at this level (corresponding to about one event per month in the Lake Erie detector) theoretical calculations indicate that events initiated by incoming neutrinos, but otherwise indistinguishable from proton decays, will start to appear. Veto counters will not detect the neutrinos, so these events cannot be separated from genuine proton decay. The limit to the proton lifetime that can be detected with this sort of detector is, then, 6×10^{32} years.

In the meantime, theorists have refined the theoretical prediction based on the simplest GUT model, outlined in Chapter 7. The current theoretical prediction is that the lifetime must be between $10^{27.3}$ and $10^{30.7}$ years. The fact that the prediction has such a large spread has to do with uncertainties connected with various technical difficulties involved in carrying out the calculations. Nevertheless, the prediction is in mild disagreement with the experimental result quoted above. Theorists are quick to point out,

Marvin Marshak's proton-disintegration detector is a thirty-ton rectangle of taconite concrete slabs and 3,456 tubes, each attached to an amplifier consisting of five integrated circuits. The amplifier enlarges any proton-decay signals. The whole unit is attached to a computer that is connected to a telephone so that results can be monitored from outside the mine where the detector is located, under two thousand feet of solid rock in northern Minnesota. Photo by Tom Foley, courtesy of the University of Minnesota News Service.

An overview of the site of Marshak's proton decay experiment. Photo by Tom Foley, courtesy of the University of Minnesota News Service.

however, that this prediction results from the simplest version of the GUT and that there are other, more complex versions, which will predict different values for the proton lifetime. As of this writing (summer 1983), the proton decay situation remains in this somewhat ambiguous state. No decays have been seen in the laboratory, but the significance of this fact has not yet been ascertained.

Summary

The GUT predicts that ordinary protons and neutrons, hitherto thought to be absolutely stable, can decay. We do not ordinarily see such processes, because the lifetime of the proton is supposed to be very long—10^{31} years or more. When the proton does decay, it changes into a positron and a meson, with the meson in turn decaying into a pair of photons. Thus, the end product of the decay is the complete disappearance of the proton.

Because no other theory has made such a prediction, the detection of proton decay has become a major goal of experimental physicists. As of this writing, no such decay has been seen, although massive experimental efforts have been made to find one. This means that the lifetime of the proton is somewhat longer than predicted by the theory, although it is relatively easy to modify the GUT to accommodate the experimental results.

PART THREE

With an understanding of the grand unification theory, we have the ability to predict the behavior of matter at very high energies. We were forced to leave our discussion of the Big Bang when we were still a millisecond away from the moment of creation, because at that point our ability to describe the interactions of particles with each other ran out. Without the grand unified theories, the next step back in time would be impossible to take. In Part Three, we take our new understanding and move into this previously uncharted territory. Our guide will be theory, for the temperatures will be so high that there is no chance that they can be produced in our laboratories in the foreseeable future. Our journey will bring us tantalizingly close to our goal of reaching the moment of creation.

Chapter 9

The Moment of Creation

It froze clear through to China.
It froze to the moon above.
At a thousand degrees below zero,
It froze my logger love.
"The Frozen Logger"
NEW ENGLAND FOLK BALLAD

The Multiple Freezings

The most important factor in the history of the universe is that it cools off as it expands. Since matter behaves very differently at different temperatures, it is not surprising that the universe has presented different appearances at different times in the past. The best analogy is a cloud of steam as its temperature drops: first we will see a gas, and then the gas will condense into drops of liquid; finally, if the ambient temperature is low enough, we will see the liquid turn into solid ice. This last change, from liquid to solid, we call freezing. For our description of the early universe, however, we will use this word in a slightly different sense. Whenever the matter in a system rearranges itself in response to a lowering of the temperature, we say that the system has undergone a freezing in the way we've used this word since Chapter 2. In this sense of the word, both the condensation of the steam into water and the transformation of water into ice would constitute freezings.

In Chapter 2 we discussed two examples of such freezings in the history of the universe. These were the formation of nuclei from elementary particles 3 minutes after the Big Bang and the formation of atoms from free electrons and nuclei some 500,000 years later. In the first stage, the temperature of the universe before the freezing was so high that any nucleus that happened to form would have been destroyed in collisions. After the 3-minute mark, the temperature of the universe was lower, so that the energy involved in collisions was less than that required to break up nuclei after they had formed. Thus, the freezing of the particles into nuclei could not have occurred until the disruptive force due to collisions was small enough to be overcome by the attractive strong interactions. A similar set of statements could be made about the formation of atoms from nuclei and electrons except for the fact that in this case the binding was due to the electromagnetic, rather than the strong, interaction. Each freezing ushered in a new era in the history of the universe, in which the predominant form of matter was different from what it had been before.

There is another aspect to a freezing besides the rearrangement of matter. When water freezes into ice, its underlying symmetry changes. In liquid water we see the same thing—randomly moving water molecules—no matter in which direction we look. In ice, on the other hand, there are certain directions in space that are different from others, a fact that is apparent when you look at a snowflake. The two effects—rearrangement of matter and change in symmetry—go together. In an ordinary freezing, like that of water into ice, there is also a difference in energy between the collection of free molecules of water at 32°F and the crystalline structure of ice at the same temperature. This energy difference must be extracted from the system to make water freeze and must be added to the system to melt ice. This is why your refrigerator requires electrical energy to make ice cubes and why those ice cubes will absorb heat from a drink. Freezings in which such an energy difference occurs are called phase transitions of the first kind by physicists. There are a few transitions in nature, the most common being that involved in the formation of an ordinary magnet, where the rearrangement of matter and the change in symmetry associated with the freezing are not accompanied by a release of energy. These are called phase transitions of the second kind. In general, the freezings that involve the appearance of new forms of matter (like atoms or nuclei) are transitions of the first kind, while those involving forces are of the second kind.

We can generalize the concept of freezing to include the

unification process by noting that when two forces become unified, the fundamental symmetry of nature changes. This is analogous to the change in symmetry in water when it freezes into ice, although the freezing of forces does not involve a fundamental rearrangement of matter such as that involved in the formation of atoms, nuclei, or ice crystals. When we speak of a force freezing, then, we shall refer to the process by which the unification of two forces is lost when the temperature drops to the point that the collision energy is less than the energy that is needed for unification.

The temperature at which forces will freeze depends on the mass of the particles being exchanged to generate the force. This, in turn, depends on the phenomenon of spontaneous symmetry-breaking discussed in Chapters 6 and 7. When the energy in collisions is much larger than the mass energy of the heavy particles involved in a unification, then the true symmetry of the force will be manifest. For example, the underlying symmetry of the electroweak interaction requires that the photon (with zero mass) and the vector bosons (with masses near 100 GeV) behave like members of the same family of massless particles. This requires a very high temperature. In the centers of stars, temperatures in our era provide collision energies less than 10 percent of 1 GeV. Even 1 millisecond after the Big Bang the temperatures were below 1 GeV. Thus, at no time in the period of history we have so far explored has the temperature been high enough to provide energies comparable to the 100-GeV unification mass for the electroweak interaction. For such temperatures, we must look in the first millisecond. The same statement is true, of course, for the much greater mass involved in the unification of the strong force with the electroweak (10^{15} GeV).

If we start at the Big Bang and move forward in time, we expect to see a series of freezings. Some of these freezings will involve the appearance of new kinds of matter, while some will involve forces. Forces that are identical at very high temperatures will undergo spontaneous symmetry-breaking when the temperature falls and will become distinct. Each freezing, whether of matter or of force, will inaugurate a new era in the history of the universe. This means that we expect the first millisecond to show the same general pattern of periods of quiescence punctuated by periods of rapid change that we have already seen occurring in the subsequent 15 billion years.

3

THE FREEZING OF THE QUARKS (10^{-4} SECOND)

As you might expect, the era of particles was ushered in by a freezing. Between 10^{-6} and 10^{-3} second (that is, between 1 microsecond and 1 millisecond after the Big Bang), the elementary particles condensed out of a sea of hot quarks. The exact time of this freezing is a matter of some theoretical debate at the moment but most theorists would put it between 10^{-4} and 10^{-5} seconds. In any case, it is clear that by the time the universe was 1 millisecond old, almost every quark had been incorporated into a familiar elementary particle. Some quarks would have combined with antiquarks to form mesons, while others would have combined with two other quarks of appropriate color charge to form baryons. From the discussion of quark confinement in Chapter 7, we know that once the quarks are incorporated into particles, they can never get out again. Theoretical calculations seem to indicate that at very high temperatures quarks will not be confined in this way, but that as the temperature falls, the forces responsible for confinement will take over and the quarks will condense into particles. This condensation can be thought of as a freezing of the quarks. The details of how (and at what temperature) this occurs are a subject of debate among theorists, but there is no doubt that there is some critical temperature above which quarks are not confined into particles and below which they will be. Thus, before 1 millisecond has elapsed after the Big Bang, free quarks might be expected to disappear from the universe.

In any process as chaotic as the early stages of the Big Bang, however, we can expect some anomalous events. There might be a few quarks that, for one reason or another, simply fail to find the appropriate partners with whom to combine. Such quarks would remain as free particles and would wander the universe like so many Flying Dutchmen searching for partners. Physicists call these particles (should they exist) relic quarks. The search for relic particles will be discussed fully in Chapter 12.

The era immediately preceding the freezing of the quarks we will call the quark era. It is characterized by the presence of free quarks, in addition to leptons and photons. In this era there is no major change in the fundamental forces. We would still see four distinct forces governing the interactions of matter, and we would have no difficulty identifying them with forces we see in our laboratories today.

THE ELECTROWEAK FREEZING (10^{-10} SECOND)

The quark era begins 10^{-10} second after the Big Bang, with the freezing of the electrical and weak forces. At 10^{-10} second the universe will have cooled off to some 10 quadrillion degrees or so. Above this temperature, there is enough energy available in interparticle collisions to create vector bosons; below this temperature there is not sufficient energy to do so. Another way of making this point is to note that before 10^{-10} second, the collision energy exceeds 100 GeV, while after 10^{-10} second it does not. In Chapter 6 we saw that 100 GeV is the energy at which the underlying unity of the electromagnetic and the weak interactions becomes manifest in nature. Thus, 10^{-10} second marks the latest time at which this unification was seen and hence the critical moment of transition at which the weak and electromagnetic forces froze out into their present form.

We shall call the period that began 10^{-35} second after the Big Bang and ended at 10^{-10} second the electroweak era. This era differs from the quark era that follows in one respect. During the electroweak era, the interactions between particles are governed by three (rather than four) fundamental forces—the strong, electroweak, and gravitational interactions.

An observer watching the process by which the electroweak force froze into the separate weak and electromagnetic forces would see the universe go from a state of high symmetry to a state of lower symmetry. He would also observe that, while there were many vector bosons being created as free particles before 10^{-10} second, this production would come to a stop. The vector mesons that were present at the time of the freezing would quickly decay, so that by the time we were well into the quark era, there would be no vector mesons as free particles. They would, of course, still play a role as exchanged particles in the weak interaction. Thus, as far as vector mesons and the weak interaction are concerned, the universe has been in its present state since it was 10^{-10} second old.

THE GUT FREEZING (10^{-35} SECOND)

In the conventional Big Bang picture, the electroweak era begins 10^{-35} second after creation. The energy available in collisions at this time was around 10^{15} GeV—the grand unification energy discussed in Chapter 7. The temperature was 10^{26} degrees Kelvin, which is almost impossible to picture. This freezing is by far the most interesting of all of the events we have discussed so far, for a number of reasons. We have known about the electroweak unification since the late 1960s, but the realization that the grand unification theory could be applied to cosmology is much more recent. Consequently, people are still excited about the idea, and it is hard not to catch some of this excitement when reading in the field. Moreover, our understanding of the GUT freezing is still so new that a lot of the details remain to be worked out. We are at the frontier of what is known about the Big Bang, and all frontiers of knowledge are exciting. Furthermore, this freezing is more complex in nature than any of the other events in the first millisecond, and much of what happens later is determined by events at 10^{-35} second. Small wonder, then, that it has received so much attention recently from physicists.

In some respects this freezing is much like the freezing at 10^{-10} second. Before 10^{-35} second the strong and electroweak forces were unified, so that there were only two fundamental forces governing the behavior of matter. The energy in the collisions of that era was so high that *X*-bosons, the particles associated with the grand unification, could be produced copiously. After 10^{-35} second, the strong force froze off from the other two, leaving us with the three forces characteristic of the electroweak era. At the same time, the temperature in the rapidly cooling universe had dropped below the point where the *X*-particles could be created in collisions. Without any way of replenishing their population, these particles decayed rapidly and disappeared from the scene, never to reappear. The understanding of their decay, as we shall see later, has been instrumental in resolving the antimatter problem discussed in Chapter 3.

If you have ever watched a lake or pond freeze over in winter, you can understand another important aspect of the freezing of the strong force at 10^{-35} second, at least as described by some versions of the GUT. An ice film does not form on a lake all at once but begins in several spots simultaneously, spreading out from these spots until it has covered the entire surface. In the theories being described, the freezing of the strong force proceeded in much the same way, with the new level of reduced symmetry being established first over small volumes in space and then growing to fill everything. When freezing occurs in this way, there are bound to be irregularities in the final result. For example, if we are talking about the formation of an ice sheet, the axes of symmetry of the crystals growing around one center will, in all probability, point in a different direction from the axes of symmetry associated with the neighboring center, as shown in Figure 40. When the ice spreading outward from the two centers comes together, there will be a region where the axis of symmetry changes from one direction to the other. The result is a pattern as shown on the right in Figure 40, where regions with different directions of symmetry fit together like pieces of a puzzle.

Each of the separate pieces of the puzzle is called a domain, and the regions where the direction of symmetry are changing are called domain walls. Domain walls are an example of the crystal defect—something that keeps the ice from being a single perfect crystal. Metallurgists have made exhaustive studies of the kinds of defects that can occur in nature, and it turns out that the domain-wall phenomenon is just one way a crystal can be flawed. Others, such as planes of cleavage in a diamond, are somewhat more complicated to describe but are familiar nonetheless.

The GUT freezing can be analyzed in the same way as a crystal. And just as every real crystal can contain defects, the freezing of the strong force can produce defects in the universe. It turns out that there are precisely three kinds of defects that can form in the universe at 10^{-35} second. Domain walls—extended two-dimensional structures in space—are one of these. The other two, which turn out to be much more interesting, are point defects and string defects.

Point defects are just what the name implies—single points in space where the symmetry changes. We show a point defect in

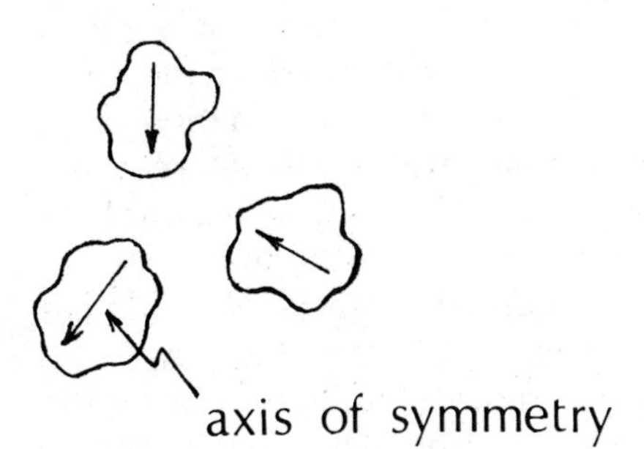

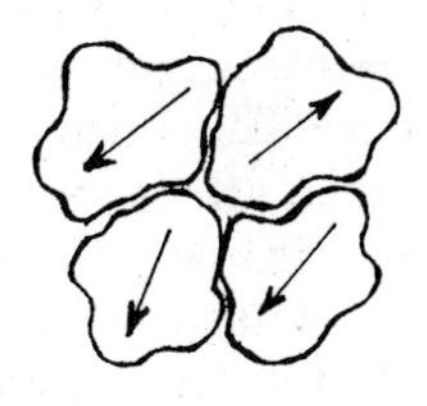

Figure 40.

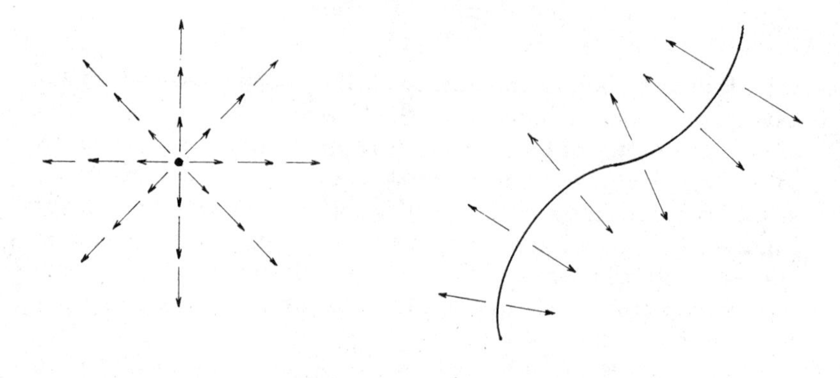

Figure 41. *Figure 42.*

Figure 41. On any line through the point, the symmetry changes abruptly. A string is a one-dimensional line separating regions where the symmetry is different. One type of string is shown in Figure 42.

When the strong force freezes, therefore, we expect that the final result should be a universe containing defects in the form of strings, points, and domain walls. The theory predicts that these defects will be very massive. (You can picture this as matter collecting around the faults during the freezing.) Point defects, for example, are supposed to have a mass of around 10^{16} GeV. To all intents and purposes, these points would appear to us to be ordinary but very massive particles. The only unusual thing about them is that they are supposed to carry an isolated magnetic charge. They would behave, in other words, like the north or south pole of a magnet. Such a particle is called a magnetic monopole, and it would be another example of a relic particle. We will discuss its significance in Chapter 12.

The string defects are also quite massive: a length of this primordial string long enough to reach across an atom would weigh a million tons. In Chapter 11, we shall see that strings may play an important role in the resolution of the galaxy formation problem. The third type of defect, domain walls, has not received much attention to date, so there is little to say about what role, if any, it may have played in the development of the universe.

Thus the GUT freezing was a complex business. Not only did the fundamental forces separate from each other, but new kinds of large-scale structures were formed in the universe during the process. At least two of these structures (points and strings) may have played an important role in later eras. Finally, this freezing marks the point after which the *X*-particles—the particles whose exchange causes the decay of the proton—can no longer be created in collisions. The *X*-particles present at the 10^{-35}-second mark decay quickly and are never seen as free particles again.

We will call the period beginning at 10^{-43} second and ending at 10^{-35} second after the Big Bang the GUT, or grand unified theory, era. This is an era of very high temperatures, with the energy of collisions ranging from 10^{19} GeV at the beginning down to 10^{15} GeV at the end. This is a hard number to visualize, but for reference the energy of motion of a fully padded fullback running at top speed is about 10^{14} GeV—ten times smaller than the energy associated with lowest temperature of the GUT era. If you imagine packing all of that energy into a collision between particles not much larger than a proton, you can get some ideas of what was going on at this time.

Because the *X*-particles could be produced copiously during the GUT era, processes in which quarks were converted into leptons and vice versa were quite common. (Some processes of this type were discussed in Chapter 8 in the context of the problem of proton decay.) This means that as particles collided with each other, their identity could be changed easily. When this situation occurs, physicists say that the interchangeable particles are actually identical, just as we would argue that changing from a jogging suit to formal attire does not affect the underlying identity of a particular individual. During the GUT era, then, there were only two kinds of particles: fermions (leptons and quarks, now understood to be identical) and bosons (the *X*-particles, gluons, vector mesons, and photons). With only two fundamental forces operating, this early universe was a very simple one compared to our own. In fact, it is a general rule that the farther back in time we go, the less complex the universe becomes.

The present state of knowledge of the GUT era is not complete by any means. Although there is general agreement about the main features that we have discussed so far, there are a number of points currently being debated among theorists. Some of these points are of a technical nature; others are more fundamental. Among the technical points is the question of exactly what kind of symmetry nature displays during this phase. In Chapter 6 we saw that symmetries play an important role in determining the behavior of the fundamental forces and that the appropriate mathematical language for describing symmetries is group theory. The jargon of group theory is esoteric, to say the least, but when physicists argue about whether the appropriate symmetry during the GUT era is *SU*(5), *SO*(10), E_7, or whatever, they are arguing about what sort of symmetry actually describes the underlying interactions. The outcome of this debate is unlikely to have much effect on the general sweep of ideas we have been presenting here.

A more fundamental problem has to do with the rate of expansion of the universe during this period. In the conventional Big Bang model, there is no radical departure from the type of expansion seen in later eras, but there are some new theories that predict something quite different. These theories suggest that there was a period when the universe expanded very rapidly. So important are these ideas that we will consider them in some detail in Chapter 10. For the moment, we note that as we approach the frontiers of knowledge, it becomes less and less possible to be dogmatic about the details of the history of the universe.

THE PLANCK TIME (10^{-43} SECOND)

The theoretical uncertainties in describing the GUT era are completely insignificant compared to what awaits us at the milestone 10^{-43} second after the Big Bang. This is called the Planck time, after Max Planck, the German physicist who was one of the pioneers of quantum mechanics.

By this time, the energy of collisions has reached 10^{19} GeV—roughly the energy of a freight train going 60 miles per hour packed into single particles. The kind of arguments that predicted the two previous unifications tell us that at this sort of energy the strength of the gravitational force should become comparable to the strength of the strong-electroweak force. If nature behaves the way we expect it to, this would be the time of the first freezing, the time when the gravitational force split off from others.

The difficulties in penetrating beyond the Planck time are prodigious, as we shall see when we discuss theoretical attempts to unify gravity with the other forces in Chapter 13. If such attempts are successful, however, the first 10^{-43} second in the life of the universe would have been an extremely simple and extremely beautiful period. There would be reactions that converted bosons to fermions and vice versa, so that there would be only one kind of particle. The unification of all four forces would leave only one basic kind of interaction. The universe would therefore show the ultimate simplicity: all the particles would be of one

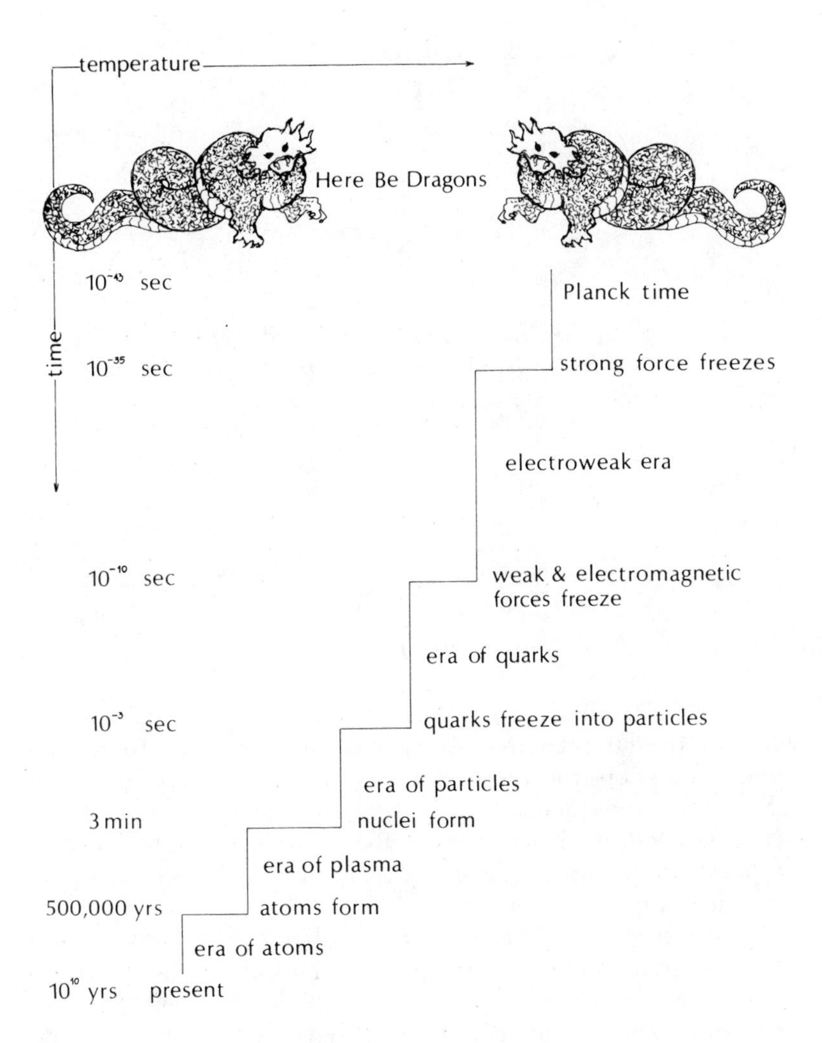

Figure 43.

type, and they would interact with each other through one kind of force. To a physicist, such a situation is so inherently beautiful and elegant that the idea simply has to be right. Whether nature feels the same way remains to be seen, of course.

In any case, our present knowledge of the early universe is summarized in Figure 43.

Chapter 10

Inflation

O if I had the wings of an angel
Over these prison walls I would fly.
APPALACHIAN FOLK SONG

Quantum Tunneling

The first major unsolved problem we encountered in Chapter 3, when we tried to trace the history of the universe back to the moment of creation, was the problem of the rate of expansion during, and just subsequent to, the GUT era. To understand what is being debated, we have to learn about one of the most unusual properties of matter predicted by quantum mechanics—the phenomenon of quantum tunneling.

If you throw a marble into a mixing bowl, as shown in Figure 44, one of two things can happen. If the marble has enough velocity to enable it to climb over the lip of the bowl, it will leave the bowl and go on its way. If it does not have enough energy to do so, it will be trapped. If there were no friction, it would continue to roll back and forth in the bowl as shown on the right.

We can represent this situation by using something physicists call an energy-level diagram. This diagram for the bowl and the marble is shown in Figure 45. The curved line, whose shape is necessarily the same as the shape of the bowl, represents the energy needed to lift the marble to that particular point against the downward force of gravity. The energy of the marble is represented by a straight horizontal line. This line is the total energy

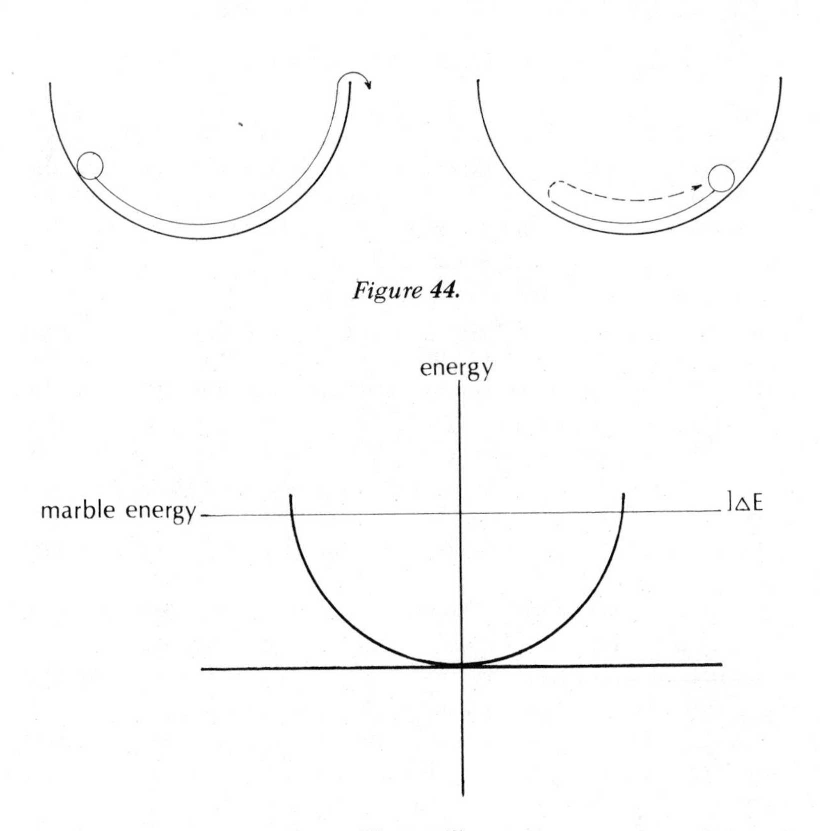

Figure 44.

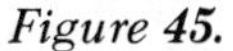

Figure 45.

that we impart to the marble when we put it in the bowl. If we think of the marble rolling back and forth, then the intersection of the line and the curve in Figure 45 represents the highest point that the marble reaches each time around. Assuming the absence of friction, the total energy of the marble does not change as it rolls back and forth; this is what is indicated when the energy is represented as a straight line in the figure. What does happen as the marble rolls back and forth is that the form in which that energy manifests itself changes. At the lowest point of the bowl it is entirely energy of motion, or kinetic energy, while at the highest point of its path it is purely energy of position, or potential energy. At intermediate points there is some mix of these two types of energies. The fact that the marble is trapped in the bowl can be deduced from the figure by noting that the horizontal line representing the energy of the marble lies below the energy required to lift the marble to the lip of the bowl.

In our experience with familiar objects, this is all we could say about the marble and the bowl. Once the marble was trapped, it would stay trapped forever. We might note that the energy of the marble is higher than the energy it would have if it were at rest on the tabletop on which the bowl sits, so that if the marble could get out, it would roll away from the bowl. This is irrelevant as long as the marble is confined.

If we replace the marble by an electron, however, our familiar concepts can no longer guide us. If physicists have learned anything, it is that there is an enormous difference between the behavior of a macroscopic object like a marble and a microscopic object like an electron. For an electron, the fact that the particle has more energy than that needed to place it on the tabletop is extremely significant, because the laws of quantum mechanics tell us that there is some probability that the electron can escape from the bowl.

This escape cannot be something that we can picture easily. For example, the laws of conservation of energy would exclude a sequence of events in which the electron rolled slowly up the sides of the bowl and fell over the lip. A more accurate visualization of the prediction of quantum mechanics would be to think of the electron leaking or tunneling through the wall of the bowl. Once through, it would find itself in an area where its energy is higher than that of its surroundings. At this point it would fall, converting its potential energy into energy of motion. It would then roll away from the bowl.

Thus, a quantum mechanical particle will not remain confined in a container forever but will eventually tunnel out. If we wanted to know how long this tunneling process would take, we would have to use the mathematical formulation of quantum mechanics to get an answer. If all we want is some notion of how such a prediction could arise from quantum mechanics, all we need is the uncertainty principle, discussed in Chapter 5. This principle, you will recall, told us that if we denote by ΔE the uncertainty in the energy of a particle and by Δt the uncertainty in the time when it will have that energy, then we must have

$$\Delta E \cdot \Delta t \geq h$$

where h is Planck's constant.

This relation tells us that it is possible for a quantum system to violate the strict law of conservation of energy, provided that the energy balance is restored in a time interval shorter than Δt. Such a process could never be detected directly, even in principle, so no laws of physics would be violated if it occurred. As far as the electron in the bowl is concerned, this means that there is some probability that it can, for short periods of time, acquire additional energy sufficient to put it over the lip. The additional energy is labeled ΔE in Figure 45. In keeping with our understanding of the uncertainty principle, we have to remember that it is not impossible for the electron to have this extra energy, provided that it has the energy for a short enough time. Since we cannot rule out the possibility that the conservation of energy can be violated on this time scale, we have to proceed on the assumption that such a process can occur. If the electron has an additional energy ΔE for a time Δt and if this time is long enough for the electron to travel over the lip of the bowl, then when the electron gives up the additional energy and returns to its original state, it will be outside the bowl. An observer watching this process would see the electron inside the bowl, and then, in a time too short to measure, he would see the electron outside. It would appear that the electron had simply tunneled through the bowl, which is the origin of the name of the phenomenon.

From this discussion, it is obvious that the ability of a particle to exhibit tunneling depends on how it is being confined. If the amount of energy needed to get over the top is very large, then the time during which the particle can possess this energy is small, and the chance of the particle being able to get out is reduced. The curve of the bowl or other restraining mechanism on an energy diagram is called the potential barrier. The higher the potential barrier, the less likely the particle is to tunnel out. Similarly, a potential that is low but wide also makes tunneling more difficult, in that it increases the distance the particle must travel in the alloted time Δt.

There are many examples of quantum tunneling in nature. Some transistors, for example, use tunneling to get electrons from one part of the device to another. Another example, somewhat easier to visualize, is the decay of some nuclei, such as uranium. These nuclei decay by emitting a particle made up of two protons and two neutrons. Known as the alpha particle, this collection is actually the nucleus of an ordinary helium atom. We can imagine the decay process as shown in Figure 46: the alpha particle is trapped in the nucleus and held there by the strong force. To

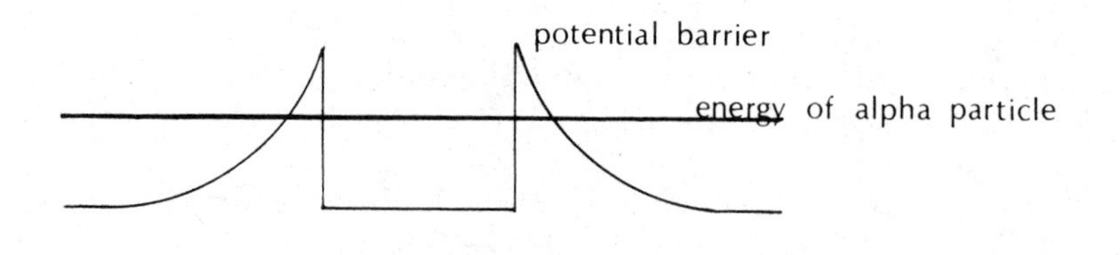

Figure 46.

escape from the uranium, its energy would have to be higher than the barrier created by that force. The analogy to the bowl and the marble is obvious.

What happens in the uranium nucleus is that the alpha particle will eventually tunnel out, converting its potential energy to energy of motion when it has done so. An observer watching the nucleus would see a spontaneous decay—the uranium nucleus suddenly emitting a fast-moving alpha particle and becoming a thorium nucleus in the process. In the case of uranium, the amount of energy needed to overcome the barrier is so high and the barrier is so broad that it is extremely unlikely that the tunneling will occur. This is why the lifetime of the uranium nucleus is several billion years. Other nuclei have lower barriers and correspondingly shorter lifetimes. We should add in passing that alpha-particle tunneling is only one of many ways that radioactive decay can occur. Other ways are similar to the proton decay discussed in Chapter 8, which occurs through the exchange of virtual particles.

Although we began our discussion of tunneling with the mechanical example of the bowl and the marble, where the potential energy associated with the barrier was purely gravitational, it should be obvious that tunneling should be ubiquitous on the quantum level and that the arguments based on the uncertainty principle can be applied to any quantum system. In the alpha decay of uranium, for example, the potential barrier that the alpha particle has to overcome to get away from the nucleus is supplied by the strong force, not by gravity. In the case of the transistor, it is electrical forces set up by atoms in the material that produce the barrier through which electrons must tunnel. At the level of quarks, it is forces between color charges that produce a potential barrier, preventing quarks from leaving elementary particles. In this case, however, the barrier has an infinite height, so quarks can never get out.

The False Vacuum

The second important concept that has to be mastered before we can discuss the rate of expansion of the early universe is that of the false vacuum. Like so much of the physics that governs the first 10^{-35} second of the universe, this is a highly abstract idea that is best understood in terms of an analogy with simpler, more familiar systems. The analog to the false vacuum that we shall use is the ordinary iron magnet.

Every atom of iron can be thought of as a tiny magnet with a north and a south pole. The magnetic field of an ordinary piece of magnetized iron, such as the one holding notes on your refrigerator door or forming the needle of your compass, is just the sum total of the magnetic fields associated with each atom of iron. A large-scale effect like this is only possible if a large number of the atoms are lined up so that their north poles are in the same direction. When this happens, we say that there is a net magnetization of the iron.

Now if we think about a piece of iron that has been heated and ask what the energy of the system looks like, we will get a graph like Figure 47. Because of the heating, any alignment of the atomic magnets will be destroyed. The atomic magnets will point in random directions and there will be no net magnetization. This is shown in the graph by the fact that the lowest energy for the system—the one toward which the laws of physics tell us the system will evolve—is the state with zero net magnetization. If we want to give the system some magnetization, then we have to reach in from the outside and line up some of the atomic magnets. In carrying out this task, we will necessarily be adding the energy needed to create the alignments, so that heated iron with a net magnetization must necessarily be at a higher total energy

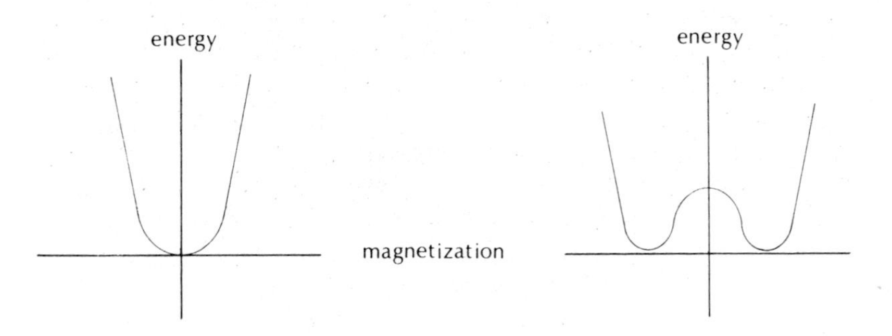

Figure 47.

than heated iron with randomly oriented atoms. Physicists think of the iron with an energy curve like the one in Figure 47 in exactly the same way that they think of the marble in the bowl; the only difference is the meaning we attach to the potential energy curve. In the example of the marble, the curve represented the energy that had to be supplied to overcome the force of gravity in order to place the marble at a certain point.

Next, suppose we lower the energy of the iron by allowing it to cool. At a temperature around 1,400°F (known as the Curie temperature), the energy curve will undergo a sudden transformation. From the smooth, bowl-shaped curve shown on the left in Figure 47, it will transform into a more irregular shape like the one shown on the right. In this case, the lowest energy state does not correspond to zero magnetization but to some value different from zero. The process behind the changing of the potential curve is simple: at the Curie temperature the interaction between the atoms is such that they tend to align themselves, a process that produces a magnetic field. No outside intervention is necessary to achieve this alignment; it happens spontaneously when the temperature is right. The transition from an unordered collection of iron atoms and the set of aligned atoms that comprise a magnet is what we have called a phase transition of the second kind. There is no overall change of energy involved in the process, just the establishing of order among the atoms in the system, with an attendant redistribution of energy between atoms and the magnetic field.

We might note in passing that this analysis of the formation of a magnet gives us another way of looking at the process of spontaneous symmetry-breaking discussed in Chapters 6 and 7. Above the Curie temperature there is a high degree of symmetry in the system. Every direction is the same in the sense that no matter in which direction we look, we see no net magnetization. Below the Curie temperature there is a preferred direction in space—the direction defined by the north and south poles of the magnet. At the Curie temperature the symmetry is broken without any interference from outside agencies—a state of affairs that explains the use of the word *spontaneous.*

Although physicists do not usually use this sort of language, there is another way we could describe the magnetized state of iron. If we were guided by our intuition about the behavior of iron above the Curie temperature, we might insist that the true ground state of iron had to be the one associated with zero magnetization. If this were the case, then we would say that below the Curie temperature iron falls into a false ground state—a state of lowest energy, to be sure, but a state that fails to meet the intuitive requirement of zero magnetization. This way of looking at things becomes important in dealing with the early universe.

In order to make the transition from a magnet to the universe during the GUT era, we have to replace the familiar magnetic field by one that is less familiar. Called the Higgs field after its originator, British physicist Peter Higgs, this field has the property of vanishing at very high temperatures, when the strong and electroweak forces are unified. Below the analog of the Curie temperature (a temperature reached by the universe at 10^{-35} second), there is a net Higgs field in the universe and the symmetry is broken. The energy curve for the universe as a function of the Higgs field is shown on the left in Figure 48 for temperatures characteristic of times before 10^{-35} second. It is similar to the corresponding curve for iron above the Curie temperature. At high temperatures we expect the universe to have zero net Higgs field, just as at high temperatures we expect iron to exhibit zero net magnetization. As the universe expands, it cools off, and below a critical temperature the energy curve turns into the one shown on the right in Figure 48. This is the analog of the curve for iron below the Curie temperature. The lowest energy state of the universe does not correspond to a zero value for the Higgs field, as it would for higher temperatures, but to some nonzero value. Just as a magnetic field appears spontaneously in iron, a Higgs field appeared spontaneously in the early universe. The universe enters the state indicated by the point labeled *A* in Figure 48, a state physicists call the false vacuum. The name arises because it is not the lowest energy state that the system could be

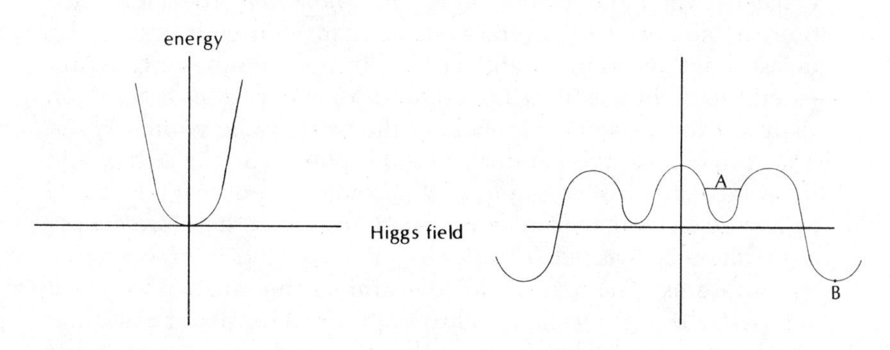

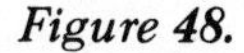
Figure 48.

in. That distinction belongs to the lower valley, labeled *B*. The universe, like the marble in the bowl, is trapped in a potential hollow at *A*, from which classical physics tells us it cannot escape to a level of lower energy, just as classical physics tells us the marble cannot get to the tabletop in our original example.

In the false vacuum, most of the energy of the universe is tied up in the Higgs field, and very little appears as matter in the ordinary sense. This is analogous to the conversion of some of the energy of the iron atoms to energy of the magnetic field in a magnet.

The Inflationary Scenario

Quantum tunneling and the false vacuum come together to produce some of the most interesting theories of the early universe currently on the scene. They go under the general heading of inflationary universe theories. The basic idea of these theories is that the universe enters a false-vacuum state like the one discussed above shortly after the Big Bang, and then tunnels out. (You can think of this as a process whereby particles in the universe tunnel out one by one.) It turns out that while the universe is in the false vacuum, it expands (inflates) much more rapidly than might be expected, which is where the name for the theories arises.

Whether or not the universe ever enters a state of false vacuum depends on whether or not the peculiar kind of dip exemplified by point *A* in Figure 48 appears in the potential-energy curve. The answer to this question depends on the details of the assumptions that theoreticians make about interactions in the universe between the Planck time and the GUT freezing. Since there is no way that interactions at such high energies can be measured, there is no "right" answer from experiment. But, provided that a theorist does not commit some egregious error, such as producing a theory that violates the basic gauge symmetry, he is free to change the parameters of his potential at will. It would be as if the theorist trying to explain the origin of magnets could never see an atom of iron or a magnet and could therefore choose any values he wished in order to describe the interaction between atoms. This paucity of constraints means that we must compare theories by their ability to produce sensible results and avoid contradicting known facts about the universe. It seems that theories that predict the presence of false vacuums satisfy this criterion.

In these theories, the universe does indeed enter the false-vacuum state. Most of the energy is tied up in creating the Higgs field. The equations of general relativity, when applied to a system of this type, predict a rather startling behavior. Instead of the normal type of expansion of the universe, in which the radius grows rather sedately as time goes by, the expansion will be much more rapid. In fact, the theory predicts that the radius of curvature of the universe will expand exponentially. The difference between normal growth and exponential growth is striking. If the expansion of the universe were normal (that is, if it were governed by the same equations at the GUT freezing as it is at later times), we might expect an eightfold increase in size during the time that the universe is in the false-vacuum state. Instead, we are confronted with a situation in which the universe doubles its size almost seventy times in an interval only a few times 10^{-35} second long. It is this hyperexpansion that is called the inflation of the universe. Theories which predict this sort of rapid expansion are called inflationary.

Just as an ordinary marble can never escape from an ordinary bowl, the universe would never be able to leave the false-vacuum state if it obeyed the laws of classical physics. We know, however, that a quantum mechanical marble could tunnel out of the bowl, and in just the same way, we imagine the universe undergoing a quantum tunneling that carries it out of the false-vacuum state. When it does so, it will find itself in a situation where its energy is higher than the potential-energy curve. Just as the marble converted this excess energy to another form, so too is this excess energy converted in the early universe. The only difference between the universe and the marble is that the former converts its excess energy into mass instead of energy of motion. In fact, in inflationary scenarios most of the mass of the universe is created when the energy of the Higgs field is "dumped" after the universe has tunneled out of the false vacuum.

While this scenario is quite different from the smooth and uniform expansion envisioned in the standard Big Bang picture, it is not unreasonable. After all, we have seen one freezing after another in the history of the universe, and we might expect that at least one of them should have an effect on the overall expansion. This is especially true of the early period we are discussing here, where high matter densities and high temperatures would surely produce effects we no longer see in our present cold and diluted era.

Thus, from our point of view, the inflationary scenario has two important consequences for our picture of the universe between the Planck time and the GUT freezing. First, it implies that the universe at the Planck time was much smaller than we would have thought if we simply extrapolated the observed universal expansion backward in time. If this were not the case, then inflation would have created a universe much larger than the one we believe existed at 10^{-35} second.

Second, the process of quantum tunneling and the subsequent creation of matter, now envisioned as a "dumping" of energy when the system emerges from the potential barrier, gives us a theoretical tool for dealing with the question of the amount of matter in the universe. Just as we can calculate how much potential energy a marble will convert into energy of motion when it tunnels through a bowl, so can we calculate the amount of energy that is converted into mass when the universe leaves the false vacuum. This is the first indication that we might be able to use the grand unified theories to deal with truly fundamental questions about the structure of the universe.

The New Inflationary Scenario

What we have described so far is a theory of the early universe in which inflation occurs because of the presence of a false vacuum and the tunneling process. In such theories, the universe inflates during the false vacuum period. As you might expect, however, this is not the only theory being examined by physicists interested in understanding the GUT freezing. In late 1982, Alan Guth of MIT, the creator of the original inflationary model, produced an alternate picture of this process, dubbed the "new inflationary scenario."

The new inflationary scenario differs from what we have discussed up to this point in two important ways. First, although there is a false vacuum, it plays a much less crucial role in the new inflationary scenario than in the old theory. Second, in the new inflationary scenario, the entire universe freezes in a single domain, so that the lumpy structure shown in Figure 40 need not form. Let us consider these two points separately.

The sort of deep hill-and-valley structure in the energy curve in Figure 48 guarantees that the universe will remain in the false vacuum for some time. In the new inflationary scenario, the postulated interactions between particles is changed so that the energy curve looks more like the one shown in Figure 49. A small valley near zero is combined with a very long, slowly falling slope to the minimum. In this case, particles tunnel out of the shallow false vacuum very quickly, but then slowly convert their energy into mass as they "roll down the energy hill." It turns out that the conditions during this slow process are such that a rapid inflation occurs, just as with the old scenario.

One of the problems with the idea of the universe freezing in separate domains is that most versions of the theories then predict that many magnetic monopoles will be created. Indeed, some predict that there should be as many monopoles as protons in the universe. In Chapter 12 we will discuss the search for monopoles in detail, but we know that if they are present in today's universe they are not present in large numbers. In the new inflationary scenario, the interactions that lead to the energy curve in Figure 49 also predict a much smoother freezing process and hence fewer defects of all kinds, including monopoles.

As we said at the beginning of this discussion, the details of

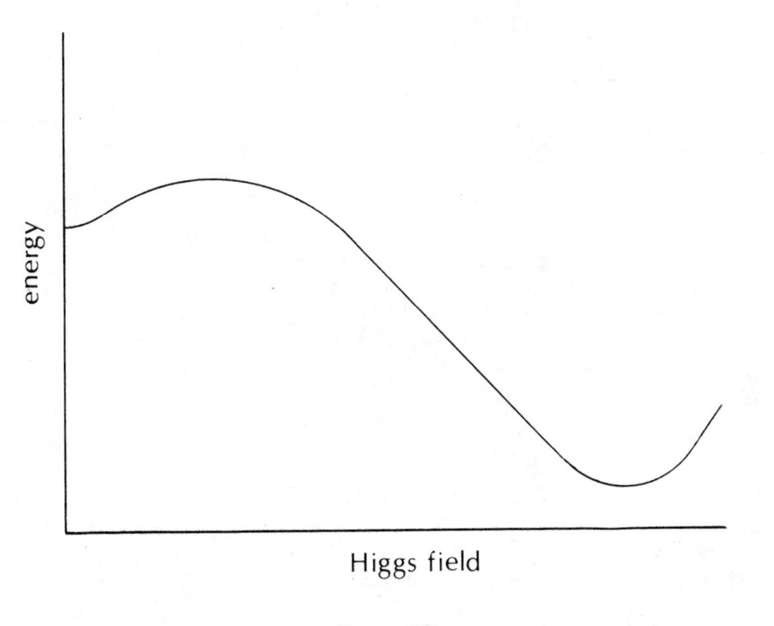

Figure 49.

how the GUT freezing occurred are still very much a subject of lively debate among scientists. I hope that this brief comparison of two separate theories will give you some idea of how this debate is conducted, as well as a feeling for what sorts of changes might occur in our theories as new ideas develop.

Chapter 11

Solving the Problems of the Big Bang

All is for the best in this best of possible worlds.
VOLTAIRE
Candide

In Chapter 3 we outlined the four basic difficulties with the standard Big Bang scenario. Unless we resort to arbitrary assumptions, it is impossible to explain (1) the absence of antimatter in the universe, (2) the formation of galaxies, (3) the isotropic nature of the cosmic background radiation, and (4) why the universe is so close to being closed. At the time, we pointed out that nothing in our knowledge of the physical laws that have operated in the universe since it was 1 millisecond old could provide us with an explanation of these phenomena. We pointed out that should such an explanation be forthcoming, it would have to involve physical processes taking place within that crucial first millisecond.

We have now studied the history of the universe as best we can back to the Planck time. We know far, far more about its evolution than we did a decade ago. It is time to reexamine those four problems in light of our new knowledge.

The Antimatter Problem

By the beginning of the particle era, there was already a preponderance of matter over antimatter in the universe. This imbalance was not as striking then as it is now. Instead of the present absence (or near absence) of antimatter, at 1 millisecond there were roughly a billion and one protons in the universe for every billion antiprotons. An observer living at that time might not have been so struck with the fundamental asymmetry as we are today, but the one leftover proton for every billion pairs of protons and antiprotons is what is responsible for all the material objects we see in the universe. The problem before us, then, is to understand how this tiny imbalance could have arisen during the successive freezings of the universe prior to the particle era.

If we begin with a universe in which matter and antimatter appear in equal amounts, there are some fundamental principles of physics that seem to tell us that we will never be able to generate an excess of matter or an excess of antimatter. Virtually all of the particle reactions we can observe in the laboratory are symmetric between matter and antimatter. Thus, if a reaction produces an extra proton, it will also produce an extra antiproton, so that the net baryon number (protons minus antiprotons) in the universe remains unchanged. If this statement were universally true, then the only way to produce the universe that we see would be to have a preonderance of matter created in the Big Bang. As it happens, however, the GUT, coupled with some very precise experimental work, show us how to get around this problem.

The reason that there was a strong tendency for physicists to believe in an aboslute symmetry between matter and antimatter has to do with something called CPT symmetry. You can think of this symmetry in the following way: Suppose you look at the film of a reaction between particles and then turn every particle into its antiparticle, turn the film over so that the right and left are interchanged, and run the film backward. The CPT theorem says that the laws of nature deduced from the film that is the final result of these operations will be indistinguishable from those derived from the original film. This is a very strong statement in physics, because every theory that obeys the principle of relativity and the basic postulates of quantum mechanics must predict this result. The three operations outlined above are called (1) charge conjugation (this is the *C* in *CPT* and corresponds to changing particles into antiparticles), (2) parity, or *P* (interchanging left and right), and (3) time reversal or *T* (running the film backward). Before 1954, physicists believed that the way CPT symmetry was achieved in nature was that each of the three operations, in and of itself, represented an unbroken symmetry. Thus, the parity operation, which interchanges left and right, was thought to imply that there was no way to distinguish between righthandedness and lefthandedness at the level of particle interactions—that all laws of physics would look the same whether seen directly or viewed in a mirror. But in 1954, two young theoretical physicists, T. D. Lee and C. N. Yang of Columbia University, pointed out in a brilliant paper that although most interactions in nature seem to occur in such a way as to keep parity unbroken, there is no reason why this must always be so. It was completely within the realm of possibility that nature was not invariant under the parity operation. All that was necessary was that if it were not, then charge conjugation, time reversal, or both would have to be violated as well. The violations associated with the various symmetries would then have to combine in just such a way as to cancel each other out and leave nature invariant under the combined CPT operation. When Lee and Yang's conjecture on the breakdown of parity was verified in the laboratory, they became (in 1957) the youngest men in history to be awarded the Nobel Prize.

For a period after the downfall of parity, physicists believed that all of the effects of parity violation would be compensated by violations in charge conjugation, so that the combined operation we have called CP would be a good symmetry. If this were true, it would mean that the laws of physics derived from watching reactions directly would be indistinguishable from those derived from watching reactions in which particles had been replaced by antiparticles and right and left had been interchanged. Alternatively, since we know CPT must be a good symmetry, this would imply that nature would have to exhibit the same laws whether we watched the film running forward or backward. If this were the case, the laws of nature would be truly symmetric between particle and antiparticle, and there would be no chance of a universe like our own developing from one that had equal amounts of matter and antimatter at the beginning.

In 1964, James Cronin and Val Fitch of Princeton University announced the results of a painstaking set of experiments which showed that at least one phenomenon in nature—the decay of a particle called the K°_{L} meson—was not invariant under the CP operation. Clearly, nature has chosen to keep CPT valid, but has arranged things so that the laws of physics are not invariant under C, P, and T taken separately. For this discovery, Cronin and Fitch shared the Nobel Prize in 1980.

This discovery was followed quickly by another that has

obvious bearing on the antimatter problem. The K°_{L} decays more often into a positron plus other particles than it does into an electron plus other particles. The K°_{L} is one of a handful of particles in nature in which the particle and antiparticle are identical. It is, in other words, its own antiparticle. If matter and antimatter are always interchangeable, then we would expect the K°_{L} to decay into matter and antimatter in exactly equal proportions. Every time a positron appears among the end products, we would expect to find another decay in which an electron appeared. The fact that this is not what happens means that here, for the first time, we see a process that does not treat matter and antimatter on an equal footing. This is heartening, because it gives us hope that we might be able to construct our theories in such a way as to generate a preponderance of matter in the early universe.

In fact, that is just what was done by a number of theorists in the late 1970s. The reasoning is actually quite simple. Before the GUT freezing, the universe was full of X-bosons and quarks. Every time an X-particle decays at one point, an X is created in a collision somewhere else. The same is true of the antiparticle, the $\bar{X}$. If the universe started out with equal amounts of matter and antimatter, nothing that happened before the GUT freezing would have destroyed the balance. Suppose, for the sake of argument, that the decay of the X produced 2 billion quarks and 1 billion antiquarks. If matter-antimatter symmetry held, we would then expect that the decay of the $\bar{X}$ would produce 1 billion quarks and 2 billion antiquarks. The net effect of the decay of an X and an $\bar{X}$, then, would be a universe containing 3 billion quarks and 3 billion antiquarks—that is, a universe with no preponderance of matter over antimatter. If the universe emerges from the GUT freezing in this state, we know of no subsequent processes that can upset the balance. Thus, if we wish to solve the antimatter problem, we must look at the GUT freezing in more detail.

Some hint of how this freezing might have occurred can be gleaned from the K°_{L} decay. We know that it is possible that the symmetry between matter and antimatter in nature can be broken. Theorists have been able to show that the same sort of processes that lead to this result in the K°_{L} decay can happen in the decay of the X-particles just after the GUT freezing. In other words, just as the decay of the K°_{L} produces more antimatter than matter, the decay of an X and an $\bar{X}$ can produce more matter than antimatter. It is not difficult to show that the small excess of matter over antimatter which we need to explain the present universe, can be easily accommodated by the GUT.

The explanation, therefore, would unfold as follows. At the time of the freezing, there were equal numbers of X and $\bar{X}$ in the universe. Immediately after the freezing, the particles could no longer be produced in collisions and began to decay. Suppose for the sake of argument that the decay of a single X particle produced a billion and one quarks along with a billion antiquarks. The theory then predicts that the corresponding decay of the $\bar{X}$ would produce a billion antiquarks and a billion quarks. In this case, the decay of the pair would have produced a small excess of matter over antimatter—just enough, in fact, to explain the observed universe. The net result of adding up all the possible ways the X and $\bar{X}$ can decay is that very early in the electroweak era the universe found itself with an excess of matter over antimatter, even though it entered the era without such an excess. The reason for this state of affairs is that the X-particles, like the K°_{L} we observe in our laboratories, decay via a process that violates the almost universal symmetry between matter and antimatter. Both these processes—the one we can measure and the one we cannot—are described by the same theory, so that we can discuss the decay of the X-bosons in the early universe without introducing ad hoc assumptions.

Once the X- and $\bar{X}$-mesons have decayed, there are no processes that can change the net baryon number of the universe. The excess of matter over antimatter is frozen into the structure of the universe. All that happens in subsequent eras is that matter and antimatter annihilate each other until all of the antimatter is gone, leaving a universe consisting entirely of matter. Thus, one of the most vexing difficulties in the conventional Big Bang model can be solved if we can understand the details of the particle processes that occurred when the universe was 10^{-35} second old.

The Galaxy Formation Problem

In order for the galaxies to form by the process of gravitational accretion we described in Chapter 3, there had to be irregularities in the universe at the beginning of the present era. These irregularities would have taken the form of regions in space where the density of matter was higher than average. Such regions would have acted as nuclei around which the galaxies could condense, much as dust particles in the air act as nuclei for the formation of raindrops. In order to have this effect, the irregularities had to be present at the most recent freezing, 500,000 years after the Big Bang.

The clumpings of matter need not have been large. In fact, for galaxies to form, it was only necessary that there be regions of space where the density is .01 percent higher than normal, a tiny concentration. For reference, a .01 percent density concentration would be analogous to a wave ⅛ inch high on a lake 100 feet deep.

The problem of galaxy formation, therefore, is not one of producing galaxies ready-made before 500,000 years. All we needed was for there to be irregularities at the .01 percent level, which were created before this time and managed to survive. The key word here is *survive.* The early universe was what physicists call a dissipative system. Local disturbances in such systems get smoothed out quickly and do not survive for long. A good analogy is to think of a single rock in a wide river. The rock creates ripples (local disturbances) in the water, but as we follow those ripples downstream, they smooth out and fade back into the general placid flow. We say that the river dissipates the disturbance.

It is hard to imagine anything as dynamic as the Big Bang proceeding smoothly and never producing local disturbances. I know of no one who has seriously argued that this is the case. What is argued, however, is that any disturbances produced in the early universe will fade out like the ripples on a river, leaving the universe without the necessary nuclei around which galaxies could form. What is needed is not so much the creation of such nuclei in the first 500,000 years but their preservation over this large time scale. In the conventional Big Bang model, such nuclei simply cannot be made by any natural process, so they have to be put in by a special assumption.

Our new understanding of the GUT freezing, however, allows us to do better than this. We know that certain types of structures were formed during this freezing and that these structures could be quite massive. From this point of view of galaxy formation, the most important of these structures were the strings—long, very massive, two-dimensional objects. Strings are one kind of defect that appeared in the GUT freezing and were discussed in Chapter 9. Theorists have traced the history of these

Figure 50.

strings from 10^{-35} second on, and have made some rather interesting discoveries. Using concepts from the branch of mathematics known as topology (the study of shapes), they have shown that strings like the ones shown in Figure 50 are stable and could be expected to survive for long periods. There may be other important types of stable strings as well; this is a brand new field in physics, and not yet fully explored.

In any case, these studies show that it is theoretically possible that defects formed during the GUT freezing could survive for the next 500,000 years and would, by that time, be just massive enough to provide the .01 percent concentrations needed to start the condensation of the galaxies. Thus, in this version of creation the galaxies form very quickly after electrons and nuclei come together to form atoms, and the difficulty posed earlier is avoided. The galaxy formation problem, like the antimatter problem, is solved by a better understanding of the behavior of the universe during its first 10^{-35} second of life.

We should note in passing that this particular explanation of galaxy formation depends on the formation of strings in the GUT freezing. It is not, however, essential that the nuclei around which galaxies condense be formed in precisely this way. In the new inflationary scenario, for example, no strings are created, but other types of mass concentrations are formed by quantum mechanical processes during the period of rapid expansion. In either case, the end result is the same—galaxies form quickly once the plasma freezes into atoms.

There is an important lesson in this result, not just for physics but for science in general. Since the early 1970s there has been a resurgence of biblical creationism in the United States. Going under the name *creation science,* this movement has produced a large body of arguments about the failings of evolutionary theory. These arguments have not been restricted to criticisms of Darwin but have been expanded to include the whole concept of the evolutionary universe as well. An important element in their argument has been the failure of conventional cosmology to solve the problem of galaxy creation.

With the development of the GUT, however, we see that galaxy formation is no longer a problem at all but simply one more natural phenomenon with a perfectly natural explanation. The creationist argument fails because it tacitly assumes that if a process cannot be explained now, it will never be explained in the future. This line of reasoning fails in the case of galaxy formation—as it does in so many other areas of creation science—because the frontiers of knowledge often move quite rapidly. It never pays to base a philosophical position on what scientists do not know.

The Horizon Problem

The observational fact that leads to the horizon problem is simple: if we look at the cosmic background radiation arriving at earth from different directions, we find that it is isotropic to one part in ten thousand. This implies that the part of the universe now above the north pole was once in communication with the part now below the south pole, so that the two parts must have established thermal equilibrium at some time in the past. The problem arises from the fact that in the conventional Big Bang picture, there was never any time in the past when this equilibrium could have been established universally, even by signals traveling at the speed of light.

We can represent this problem as in Figure 51. We plot the

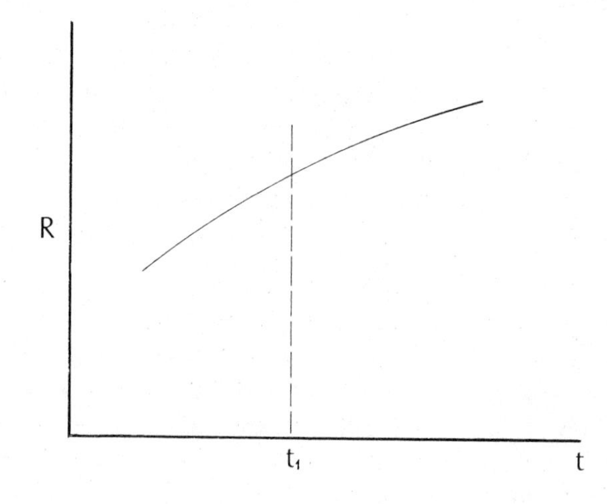

Figure 51.

radius of curvature of the early universe as a function of time. The universal expansion is represented in this graph by the increase in the radius as time goes on. At any time in the past (for example, the time labeled t_1 in Figure 51), we can find the radius of curvature from the graph. If we use this radius as a rough measure of the size of the universe at the time t_1, it is relatively simple to discover whether the radius at time t_1 is small enough so that light can travel from any point in the universe to another before the expansion moves the target point too far away. In terms of this graph, the horizon problem can be stated as follows: In the early universe, the radius of curvature was too large to allow communication between distant regions. In other words, the universe was always too large for equilibrium to be established, even in the early stages just after the Big Bang.

But in Chapter 10 we saw that the sort of smooth increase of the radius with time that is shown on the graph need not be what actually happened. The GUT predictions allow us to talk of an inflationary period at 10^{-35} second, a period when the expansion was much more rapid than that shown on the above figure. In fact, the radius of curvature as a function of time for the inflationary universe is shown in Figure 52. Before and after the inflationary period, the radius grows smoothly. During inflation, however, the growth is very rapid. For reference, we show by a dotted line the radius for the conventional Big Bang expansion.

If we now look at the time t_1, we see that the radius is much smaller in the inflationary universe than it would be in the

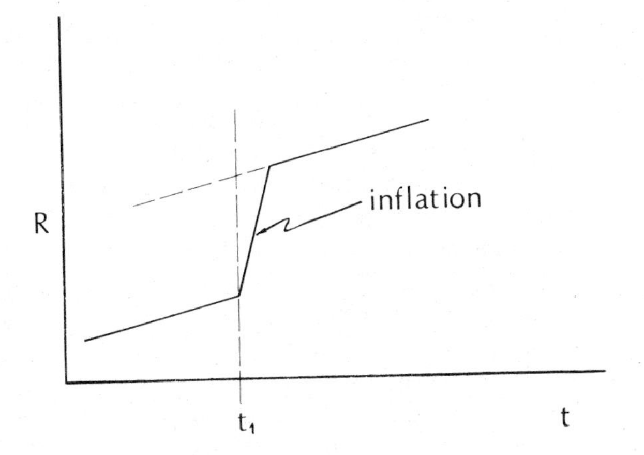

Figure 52.

normal expansion. This means that if we repeat our speed-of-light analysis for a universe that undergoes inflation, we find that before the inflationary period it is indeed possible for every part of the universe to establish communication with every other part. Therefore, the universe could very well have been in thermal equilibrium before the inflation started and distances became too great.

The solution to the horizon problem that arises from the inflationary model is thus quite easy to visualize. Sometime before the GUT freezing, when the universe was very small, thermal equilibrium was established. Parts of the universe now above the earth's north pole were in contact with parts now above the south pole. Then, when the universe was roughly 10^{-35} second old, inflation occurred. The universe expanded rapidly, but every part carried with it the memory of an earlier era—an era when everything was at the same temperature. We see the results of this memory when we measure the microwave background. We find no section of the universe where the radiation differs greatly from that associated with any other part. The microwave radiation is typical of that emitted by a body that is the same temperature throughout, and the reason for this is that once, in the very distant past, the universe was in thermal equilibrium.

Perhaps a more mundane analogy will help in visualizing the role of inflation in the early universe. Suppose we had two models of a dog, one that grew in size at the same rate we would expect a puppy to grow and another that was actually a balloon which could be rapidly inflated. If we saw the two models when they had achieved their full size, we would probably say that one had been about as big as the other throughout the growth period. This guess would be reasonable, but it would be wrong if the balloon-type model had been inflated at some time in the past. For the sake of definiteness, suppose both models are now the size of a 6-month-old dog and that the inflation occurred 3 months ago. Then the size of the balloon model prior to the 3-month mark would be much less than that of the normal model. Any argument based on a "normal" growth expectation for the balloon model in the first 3 months of its existence would be wrong, since the size of the model during that period was much less than normal.

In the same way, the inflationary universe was much smaller than normal before 10^{-35} second had elapsed. In its small state, it was possible for signals to travel to all of its parts and for thermal equilibrium to be established. Then inflation occurred, and the size of the universe became "normal." The horizon problem came up because we have assumed we could extrapolate the size of the universe back without taking the brief but important inflationary period into account

The sheer magnitude of the effects of inflation on the universe can be realized if we note that most inflationary theories predict that sometime around the GUT freezing the radius of curvature of the universe increased from about 10^{-23} cm (10 billion times smaller than the size of a single proton) to something around 10 cm—the size of a grapefruit. Since this is an expansion of space itself, rather than a movement of matter through space, inflation does not violate the strictures of relativity against faster-than-light travel.

The Flatness Problem

The flatness problem is not a problem in the same sense as the antimatter, galaxy formation, and horizon problems. Instead, it is a question about why the universe is constructed the way it is rather than some other way. In particular, the flatness problem has to do with the rather extraordinary fact that the amount of matter in the universe is close to the amount that is needed to stop the universal expansion through the action of mutual gravitational attraction.

This is not a question that must be answered in order for the Big Bang model to survive. There is no compelling reason why the universe should have this mass, but there is also no compelling reason why it should not. Thus, finding a solution to this problem is less crucial to the development of cosmology than was solving the others. Nevertheless, the inflationary model provides, as a sort of bonus, some indication of how the flatness problem might be resolved.

We saw that the final outcome of the inflationary process in the early universe was the tunneling of the universe out of the false vacuum. One of the results of this tunneling was the conversion of the excess energy of the universe into mass. If we take this idea seriously, then, our theories ought to tell us how much mass was created at this time.

Such a prediction would result from predicting the shape of the potential curve, the energy level of the false vacuum, and the energy difference between the false vacuum and the final state of the universe—the quantity we called ΔE in Chapter 10. Knowing how much energy is available for conversion, it is relatively straightforward to calculate how much matter there should be in our present era. This number is a by-product of the same calculations that need to be done to describe the inflation process itself.

While this is a very new field, and all statements about it must be taken with a large grain of salt, theorists who have produced numbers for the mass of the universe in this way have generally come up with predictions between .01 and 10 times the critical mass needed to close the universe. Thus, it appears at present that if we adjust the potential barrier so that the GUT freezing has the right properties, the eventual mass of the universe comes out right as well. It is a nice example of serendipity, and it also reduces the arbitrariness of our description of the universe.

Chapter 12

The Search for Relic Particles

Everybody in New York is looking for something.
Every once in a while, somebody finds it.
DONALD WESTLAKE
The Dancing Aztec

The early history of the Big Bang seen from the point of view of the GUT has most of the attributes we expect of a great scientific theory. It is beautiful, elegant, and intellectually satisfying, and it gives us insights into the workings of nature that no previous theory has been able to provide. Only one ingredient is missing from this list of attributes. We demand that every scientific theory be buttressed by a mass of direct experimental confirmation, and there is precious little of that in what we have been discussing for the past few chapters.

That this should be the case is not surprising. We have seen that the most crucial period in the early universe—the one that determined much of the later course of evolution—was the GUT freezing, 10^{-35} second after the Big Bang. The collision energies characteristic of that time were so high there that is no chance that we can produce them in our laboratories, at least not in the foreseeable future. Consequently, we are prevented from performing direct laboratory tests of the processes that occurred at that transition. The classical sequence of prediction, direct test, and confirmation to which physicists are accustomed simply cannot be done in this case. We are reduced, therefore, to testing our theories of the early universe indirectly.

We already have seen one good example of such an indirect test when we discussed the discovery of cosmic microwave radiation. In that case, we saw that the release of radiation that followed the formation of atoms 500,000 years after the Big Bang could be detected today in the all encompassing microwave radiation that bathes the earth. One way of thinking about the radiation is to regard it as an echo of the Big Bang—something that was created long ago but that has not disappeared and is therefore still available for us to detect. In the same way, we can look for relics of the early universe to test GUT theories. As it happens, there is a long history of searches for two such relics, although these searches were motivated by more general considerations than those involved with testing ideas about the early universe. The relics involved are quarks and magnetic monopoles.

Quarks were supposed to have disappeared from the universe when they froze into the elementary particles 1 millisecond after the Big Bang. We would expect, however, that some quarks would not have been near suitable partners and so missed the chance to combine. These quarks would still be present in the universe, and if they were abundant enough, it might be possible to detect them in our laboratories. A positive result in the quark search would lend credence to the idea of the composite nature of elementary particles, and thus bolster our faith in quantum chromodynamics. Since this theory is an important element in the grand unification, the discovery of a free quark would provide us with some experimental evidence, albeit indirect, that our vision of the early universe as a series of successive freezings is correct.

The negative of this statement, however, is not necessarily true. In Chapter 7 we saw that the GUT predicts that in our present era, quarks ought to be confined inside elementary particles. Thus, the failure to find free quarks would not cast doubt on any current theory. All that would be necessary to explain this outcome would be to argue that the number of relic particles resulting from the combination process at 1 millisecond is too low to be detected in a particular experiment. There is no theoretical indication as to the number of relic quarks that ought to be present, so this escape route is always open.

The situation is slightly different with magnetic monopoles. As we have seen, the GUT freezing is supposed to have produced these particles by the same sort of process that produced strings. How many monopoles were created, and what their subsequent behavior has been, are topics of intense debate among theoreticians at the present time. Nevertheless, the detection of a particle carrying an isolated magnetic charge (the equivalent of a north or south pole of a magnet), with a mass characteristic of the unification mass (10^{16} GeV or more), would certainly be strong evidence that something like the GUT freezing actually occurred in more or less the way we have described.

As was the case with relic quarks, the failure to find a monopole would not necessarily disprove the grand unification picture of the early universe. For monopoles, however, a continued failure in the search would require a convincing theoretical explanation. The particles are supposed to exist, and if they are not found, there had better be a good reason.

One of the more intersting coincidences in the search for relic particles is that the best candiates for the discovery of quarks and magnetic monopoles are products of the same laboratory (although they were made by different investigators). The Low Temperature Laboratory at Stanford University is under the direction of William Fairbank, a man who has devoted much effort to the quark search since the late 1960s. A young investigator in this laboratory, Blas Cabrera, stunned the scientific world in 1982 when he announced the recording of an event that might very well be the first detection of a magnetic monopole. Both Fairbank's quark and Cabrera's monopole have been the subject of much debate in the scientific community, and it is fair to say that neither has yet been accepted without question. On the other hand, neither has been definitely rejected. In what follows we will discuss the background and current status of the search for both types of relic particles.

Magnetic Monopoles

During the GUT freezing, defects formed in the universe as the freezing progressed. (These defects were discussed in detail in Chapter 9.) The point defects have been described as microscopic "twists in space," places where the direction of symmetry abruptly changed. They would be analogous to the defect in a magnetic field of the type shown on the left in Figure 53, a point where the magnetic field changes discontinuously on any radial line.

The point defects in the early universe can be examined theoretically. For example, if we draw an imaginary surface around a

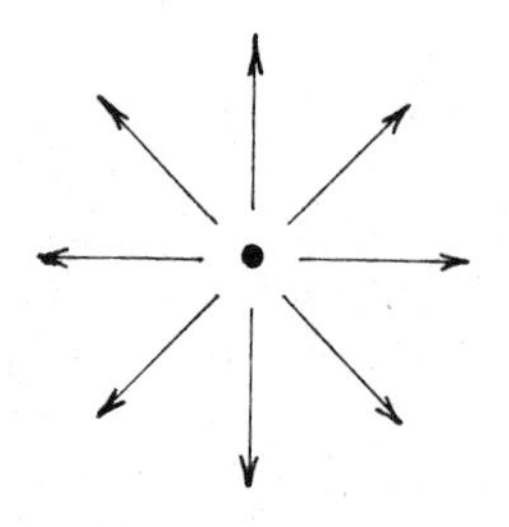
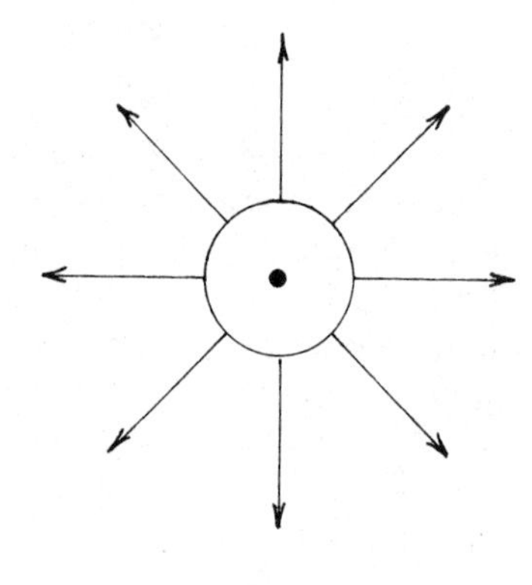

Figure 53.

defect, as shown on the right in Figure 53, we can ask what the theory predicts about the magnetic field at points on the surface. It turns out that if the surface is far enough away from the defect, the magnetic field points radially outward. Furthermore, the field has exactly the strength it would have if an isolated magnetic pole were located at the position of the defect. From this calculation, theorists conclude that defects formed during the GUT freezing would appear to us as particles producing the same effects as an isolated magnetic pole. The existence of such particles were postulated almost a century ago, and they are called magnetic monopoles.

There are a couple of points in this discussion that probably need elaboration. What, for example, is the connection between something called a defect and an elementary particle? A good analogy to the way physicists think about this question is to imagine a long line of cars in a traffic jam. If the lead car moves forward a few feet, the next car will move up to fill the space, the third car in line will then move forward to fill the space vacated by the second car, and so on. We could describe this process as we have above, totally in terms of the motions of the cars in the line. It would be equally correct, however, to describe the situation in terms of the motion of the gap. We could, for example, say that a gap had appeared in the line and moved backward at the rate of so many feet per second. The gap is actually a defect in the line of cars, but for all intents and purposes, it can be thought of as an entity having an existence of its own. In a similar way, the point defect produced during the GUT freezing would look to us like a particle—a particle with unusual properties, of course, but a particle nonetheless.

The second point that needs clarification has to do with the strength of the magnetic field produced by the magnetic monopole. Asking about this field is equivalent to asking how strongly one magnetic monopole will attract or repel another. At first glance, it might appear that the strength of the magnetic interaction, like the strength of the electrical charge, might be one of those quantities in nature that we can measure but which we cannot calculate. In 1931, however, British Nobel laureate P. A. M. Dirac showed that this was not the case. Using arguments based on the general principles of quantum mechanics, he showed that if we added the interaction between magnetic monopoles to the table of fundamental forces given on page 75, its strength would be at least sixty times that of the strong interaction. It would, in other words, represent the strongest force in nature.

Because of Dirac's finding, it has always been assumed that if a magnetic monopole were to pass through an experimental apparatus, it would leave behind it unmistakable evidence of its presence. The electromagnetic force associated with the moving monopole would act on electrons in atoms, tearing them away from their nuclei. The result would be a trail of disrupted atoms that would be quite easy to detect. Since atoms from which electrons have been removed are called ions, the process by which a monopole might remove electrons from atoms is called ionization.

Until a few years ago, all searches for magnetic monopoles depended in one way or another on the idea that they would ionize heavily in matter. Either the experiments were set up to detect directly the debris that the monopole was expected to leave in its wake, or it was assumed that a monopole striking the earth would lose energy so rapidly through the ionization process that it would slow down and stop. In the latter case, the monopole would be like a rock thrown into a jar of molasses: it would never get very far from the point at which it entered the earth. The logic of experiments designed to exploit this behavior is to search in areas where monopoles might be expected to congregate—near the magnetic poles, for example, or in moon rocks. All searches, both direct and geological, have been unsuccessful.

The development of the GUT has led to an explanation of this puzzling lack of success. Until recently, physicists believed that magnetic monopoles would have a mass of a few GeV. There was no firm justification for this belief, but there were no known particles more massive than this, and it simply never occurred to anyone that a monopole would be radically different from other particles. A particle carrying a magnetic charge will be accelerated in a magnetic field, just as particles carrying electrical charges are accelerated in electrical fields. There is a small magnetic field that extends throughout the galaxy. Its origins are not well understood, but its effects can easily be measured. The galactic field is small—about a millionth the strength of the earth's field—but it extends over huge distances. A magnetic monopole with a mass of only a few GeV would be accelerated quickly in the galactic field and would approach the earth at almost the velocity of light. A monopole moving that fast would definitely cause a great deal of ionization. The lack of success in the experimental searches caused many researchers (the author included) to conclude that there were very few monopoles in the universe, and perhaps none. Why this should have been so remained puzzling, but it seemed to be a firmly established fact.

The GUT monopoles, with their huge mass, changed this picture completely. The galactic field will indeed accelerate a heavy monopole, but it will not be able to impart a very high speed. GUT monopoles can be expected to approach the earth at .1 percent of the speed of light or less—a snail's pace compared to other cosmic rays. At this velocity, monopoles will not cause much disruption in atoms, primarily because they go by so slowly that atoms have time to adjust to the forces being exerted on them. In just the same way, a bullet will tear through a door, but if you push on the door with your hand, it will open slowly, remaining undamaged all the while. In other words, all of the monopole searches based on the idea that monopoles will cause a great deal of ionization would not have detected GUT monopoles at all. It was a whole new ball game.

The shift in thinking about monopole searches took place in a remarkably short time. Perhaps a personal experience will make this point. I wrote a paper showing that the massive monopoles would move slowly and ionize very little and submitted it to a prestigious American physics journal, where it was rejected on the grounds that it had been "shown" that monopoles would cause heavy ionization. After some minor revisions, I submitted the paper to another journal. By this time the ideas on monopoles had changed so much that it was accepted without comment. By the time it appeared in print, four months later, it was old hat. By then everyone knew that massive monopoles could not have been detected by previous monopole searches. A scientific paradigm had undergone a massive change in a period of less than six months.

But if we cannot detect monopoles by seeing the effects of their interactions with atoms, how can we detect them? It turns out that there is another important property that any magnetic pole has. If there is a loop of wire arranged so that the monopole's path takes it through the loop, then the passage of the monopole will cause an electrical current to flow in the loop. If the loop is made of ordinary copper, then collisions between atoms of copper and the moving electrons that constitute the current will eventually cause the current to subside. If, however, the loop is made of a special type of material called a superconductor, the current will not disappear but will go on flowing indefinitely. The current flowing in the loop, then, is a permanent record of the passage of a magnetic monopole.

The problem with this detection scheme is that materials have to be kept at very low temperatures if they are to be superconducting. In a typical experiment, low means between 1° and 10° above absolute zero (a number that translates to a temperature between −263° and −272° Celsius). This means that searching for monopoles by measuring their effect on a loop of wire requires rather special techniques—the techniques associated with the field we call low-temperature physics. This explains, at least in part, why the event that is the most likely candidate for the discovery of a monopole was not made in a laboratory devoted to the study of high-energy physics.

Blas Cabrera had designed an experiment to measure one of the fundamental constants of physics, an end that could be most easily achieved by working at low temperatures. Part of the apparatus for his experiment consisted of a loop of superconducting metal measuring a few inches across. He realized that if he recorded the current in this loop when the rest of the experiment was not running, the loop would act as a monopole detector. Thus, although the experiment was not designed expressly to detect monopoles, a monopole search could be carried out as a side benefit. The loop would simply sit in the apparatus, and if a monopole happened to fall through it, the event would be recorded.

The monopole search ran for six months in the fall and winter of 1981–82 without any clear indication of success. Then, at 1:45 P.M. on Valentine's Day, 1982, with no one in the lab, it happened. The current through the loop jumped by just the amount

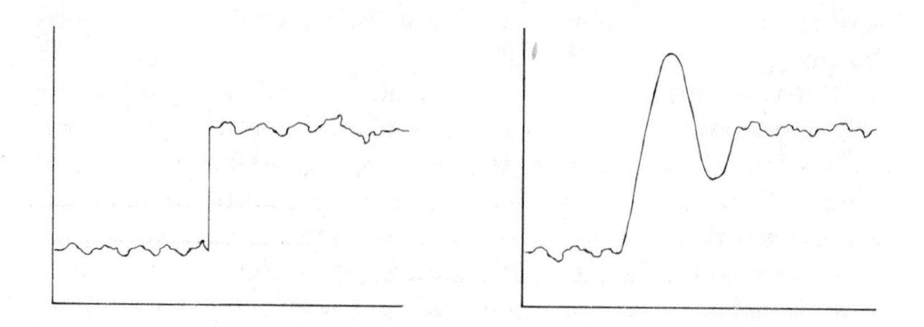

Figure 54.

that it would if a monopole had passed through. The current in the loop as recorded on a roll of graph paper, had the appearance shown on the left in Figure 54. The sudden jump in the current is exactly what would be expected from the passage of a monopole. The change in current, which is proportional to the height of the jump, is exactly right, too.

Despite this seemingly convincing evidence, Cabrera's first reaction was to ask himself, "What has gone wrong?" From his experience in the laboratory, he knew that unexpected events show up frequently in any system, so that some caution has to be exercised in interpreting such an isolated occurrence. Cabrera toyed with the idea that someone had come into the lab to borrow some equipment and bumped into the apparatus. "After all," he said, "such a person might now be too embarrassed to tell us about it." To check this possible source of the event, Cabrera recorded the current in the loop while he tapped the delicate apparatus with a screwdriver. The signal from this sort of jostling is shown on the right in Figure 54. It is clearly different from the Valentine's Day signal. By a process of elimination involving many such tests, Cabrera was able to rule out most alternative explanations for the event, leaving the passage of a magnetic monopole as the most probable interpretation of the event.

From the fact that he saw a monopole signal in his small loop after 185 days, Cabrera concluded that there must be roughly six monopoles going through each square meter of the earth's surface every day. That would mean that between five and ten go through your body each day. (It's a good thing they are moving slowly and cannot cause much damage!)

The announcement of this event touched off a wave of excitement among physicists, and many advanced monopole-detection experiments are being built at the present time. Cabrera had an improved system on the drawing boards when he made his discovery, and it went into operation in the fall of 1982. As of the spring of 1983, he has not seen a second monopole, which means that his estimate of the number of monopoles in the universe has to be reduced by a factor of at least 10—that is, instead of 6 per square meter per day, it can be no more than .6.

If monopoles are indeed rare, this kind of result is not too surprising. You can be lucky and see one soon after you start searching and then not find another for years. The longer you have to wait for the second event, the more rare the monopoles are. Anyone who makes a hobby of trying to find relatively rare things (book collectors, for example) is familiar with this fact. Nevertheless, the experimental result is so clean and the analysis by Cabrera is so rigorous that most physicists feel that there is a very good chance that this event was the first sighting of a magnetic monopole.

If this is the correct interpretation, it would be very important for our ideas about the early universe.The monopole would be a visitor that has come to us directly from the GUT freezing, and by studying the properties of the particles and their abundance, we can refine our understanding of that event.

There are, however, some difficulties with this interpretation of Cabrera's result that have nothing to do with his experiment. We expect that monopoles, having been created in the early stages of the Big Bang, should be similar to the microwave radiation in that they ought to be equally abundant everywhere in the universe. If a monopole could be expected to cross a square meter on the earth's surface every three days (or every ten days or every month), then we would, on the basis of this argument, conclude that they ought to cross a square meter anywhere in the galaxy at the same rate. In other words, our first impression is that we can simply extrapolate the monopole flux found at the earth's surface to places far distant from us.

If we assume that monopoles in our own galaxy are as abundant as Cabrera's event seems to indicate, some theoretical problems arise. One of these has to do with the existence of the galactic magnetic field to which we alluded earlier. Monopoles would be accelerated by this field, so that energy would be transferred from the magnetic field into the energy of motion of the monopoles. The effect of this transfer would be to diminish the strength of the field. If the energy taken from the field by the monopoles were not replaced and if the monopoles are as abundant as the experiment seems to indicate, then the galactic field would be drained of its energy in a short time—perhaps a million years or so. Furthermore, our best theoretical notions of the origins of the galactic magnetic field indicate that it would be impossible to establish the field in the first place if there were an abundance of magnetic monopoles. The drain on the energy of the field would be too great; it would be like trying to move a car with the brakes on. The fact that there is a seemingly permanent magnetic field in the galaxy argues that the number of monopoles crossing a square meter on the surface of the earth each day ought to be less than .01 (that is, there ought to be not more than 1 every hundred days). This is a factor of 600 higher than the number derived from Cabrera's original finding, but only a factor of 60 higher than the limit Cabrera would set as of this writing. The limit on the number of monopoles derived from the galactic magnetic field is known as the Parker bound, after University of Chicago astrophysicist Eugene Parker, who first discovered it.

Another limit on the number of monopoles in the galaxy may come from recent calculations in the GUT theory itself. Because we can understand the properties of monopoles by using the GUT, we can predict what will happen to monopoles when they encounter large collections of matter.

It turns out that GUT monopoles may have the ability to catalyze the decay of protons—that is, protons near such a monopole might be more likely to decay than those sitting by themselves. In order for this process to have any large-scale effect, the monopole has to have a close encounter with a large number of protons (or neutrons), which means that the catalysis will be most important in very dense systems.

There are in the galaxy many systems in which matter exists in states of very high density. A neutron star, for example, might be only 10 miles across and yet have the same mass as the sun. A monopole moving through a neutron star would have many close encounters and could therefore be expected to cause many decays of the neutrons. If there were as many monopoles in the galaxy as Cabrera's results indicate, and if these monopoles do indeed catalyze decay, then we would expect the extra energy added to neutron stars by the induced decays to cause those stars to give off large amounts of X rays. From the results obtained from X-ray telescopes in satellites, we know that this is not the case. Although there is a good deal of theoretical debate on the details of the calculations involved, there is general agreement that the observed properties of neutron stars are inconsistent with the Cabrera result, perhaps by a factor of a trillion or more. In other words, some theorists claim that the actual number of monopoles in the galaxy has to be a trillion times lower than Cabrera's estimate.

This conclusion is not universally accepted. While acknowledging that monopoles coming into very close contact with protons will indeed catalyze decay, another school of theorists argues that the intense electromagnetic forces associated with the approach of a monopole will push protons out of its path before the catalytic process can get under way. The calculations that have to be done to resolve this issue are immensely complicated, and it may be some time before they are done. The possible cataylsis of proton decay by monopoles is likely to be a subject of debate for some time.

From these theoretical arguments, we come to the conclusion that the debate about the Cabrera event must have one of the following outcomes:

1. The experiment is either wrong (unlikely), or Cabrera was extraordinarily lucky in observing a very rare monopole passage. In this case no more monopoles will be detected for a long time.
2. The theorists have overlooked effects that would weaken or negate their conclusions (this has happened before in similar situations).
3. Monopoles are not spread uniformly throughout the galaxy.

The last alternative has been pushed strongly by Sheldon Glashow, one of the founders of the unification theory. He argues that because monopoles are so heavy, they will interact with other matter through the gravitational force and will therefore tend to be concentrated in regions of high matter density. In this case, we could have our cake and eat it too: the density of monopoles in the region of the earth could be high enough to measure, while at the same time there would be very few monopoles in interstellar space to destroy the galactic magnetic field and the neutron stars.

The Search for Relic Quarks

The search for quarks has not been going on as long as the search for monopoles, but it has been considerably more intense (until lately, at any rate).* The result of the search can be summarized in one sentence: With one exception, the direct and geological searches for quarks have not produced any positive results.

Until the theory of quark confinement (see Chapter 7) was developed, this fact posed rather severe conceptual problems. How could quarks be the basic building blocks of matter and still not be seen in the laboratory? It would be like talking about protons and neutrons without ever seeing one in isolation. The confinement theory tells us, however, that in our present cool era, we should not expect to see quarks in isolation but only in color-free combinations. This result has taken some of the urgency out of the quark search, although it still is reasonable to talk of finding relic quarks that, for one reason or another, were left out of the quark freezing.

William Fairbank has long been involved in the quark search, and his apparatus, located in the same building as Cabrera's, has produced the only surviving candidate for the actual detection of a quark. Figure 55 is a simplified sketch of the Stanford quark search apparatus. The logic of the experiment is easy to write down, although it is technically quite difficult to carry out.

First, a niobium sphere .25 millimeter across—about the size of a dust mote—is placed in the magnetic field between two glass plates. At temperatures near absolute zero, the magnetic forces on the ball push it up, a process somewhat similar to that by which an ordinary magnet picks up a nail. If the field is adjusted correctly, the upward magnetic force can be made to cancel out the downward force of gravity, and the ball will float in space, levitated by the magnets.

Once this balancing act has been done, the total electrical charge on the ball can be measured by applying voltage to the glass plates and watching the ball move in response. In practice, the Stanford group does not measure the charge directly but tries to reduce the charge on the ball to zero. For example, if there is a positive charge on the ball (perhaps some electrons got scraped

*The history of these searches is given in Chapter 11 of my book *From Atoms to Quarks* (Scribners, 1980).

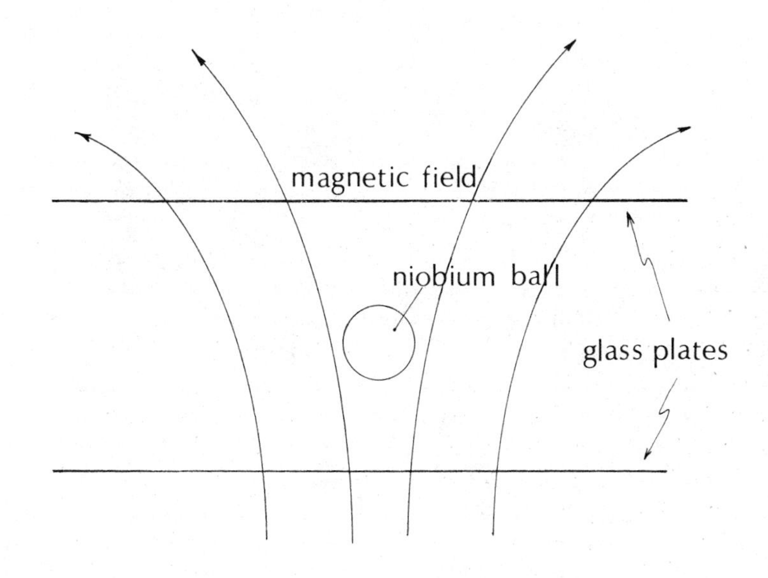

Figure 55.

off in handling), negative charge is sprayed on to cancel it. A similar procedure is followed if the ball has an initial negative charge. The end result, then, is that the charge of the ball is taken as close to zero as it can be made to go.

If there are only ordinary protons and electrons on the ball, this procedure will eventually produce a ball with zero total charge. In this case, the ball will not respond at all to the voltage on the glass plates. If, however, there is somewhere in the ball a fractionally charged particle, it will be impossible to get the charge to zero by adding single units of positive or electrical charge. The total charge will always be $\pm\frac{1}{3}$ or $\pm\frac{2}{3}$. A ball that behaves in this way is then said to contain a quark.

There is a useful analog to this experiment. Suppose you had a bank account where you were only allowed to pay or withdraw your money in even dollar amounts. If you had a positive balance, you could start withdrawing dollar bills to reduce it. If the balance became zero at some point, then you would know you had started with an even dollar amount in the account. But if you ever reached a stage where withdrawing a dollar sent the balance negative and replacing the dollar made it positive again, you might use this as evidence for the existence of fractional dollars (quarters, for example).

Fairbank ran the first version of this experiment in 1969 and found a niobium ball with a charge of $-.32$, almost exactly a third of the charge on the electron. This, of course, is what one would expect if there was a quark attached to the niobium in some way. The sources of error in the experiment were such, however, that he decided not to publish this result. Instead, he spent a good part of the next decade rebuilding and improving his apparatus. Running the improved experiment in the mid-1970s, he found that fractional charges of one-third and two-thirds were left on a number of niobium balls, a result he announced in 1977. Since that time the number of fractionally charged objects measured in the Stanford Laboratory has climbed above a dozen. Despite the subsequent years of debate and intense scrutiny of the experimental technique, this result stubbornly refuses to go away.

At the same time, no one else has conducted a successful quark search, although a number of groups around the world have been working very hard to do so. This leaves the status of the relic quark in limbo. The experiment is too well done and too rigorously analyzed to be ignored, but the lack of independent confirmation means it cannot be accepted without question either.

Summary

The status of the searches for relic particles created early in the Big Bang is somewhat ambiguous. According to some versions of the GUT, magnetic monopoles should have been created fairly copiously 10^{-35} *second after creation and should still be in evidence around us. There is one strong candidate for the detection of a monopole, but theoretical considerations suggest that they may be less abundant in our galaxy than this experiment might indicate. Until another monopole is seen, there is no way to provide an absolutely firm interpretation of the Cabrera event.*

The quark search provides the same experimental picture. The evidence for the existence of fractionally charged particles cannot be explained away but has yet to be reproduced independently. This search is less crucial as far as the theories of the early universe are concerned, since it is always possible to arrange the theories so that the abundance of relic quarks is undetectably small. Nevertheless, a positive outcome of the search would give us added confidence that our concept of the elementary particles as composite structures is correct.

Chapter 13

Beyond the Planck Time: Asking the Ultimate Questions

The answer to life, the universe, and everything . . . is 42.

DOUG ADAMS
Hitchhiker's Guide to the Galaxy

The Problem

When a final version of the GUT is developed, we will be able to trace the universe back to the Planck time, within 10^{-43} second of the Big Bang. The significance of this time was first recognized by Max Planck, the German physicist who was one of the founders of quantum mechanics. In 10^{-43} second, light can travel 3×10^{-32} cm—less than a quadrillionth of the distance across a proton. Quantum mechanics tells us that a particle that is to probe distances this small must have a very large energy—an energy comparable to the gravitational energy of two particles 3×10^{-32} cm apart. From this we conclude that on this scale of time and distance, we cannot neglect the effects of quantum mechanics on the force of gravity. If we wish to go beyond the Planck time, we will have to have at our disposal a quantum mechanical theory of gravitation. No such theory exists at the present time, but an intense theoretical effort is under way to find one.

If such a theory were developed, we would be able to push our knowledge of the history of the early universe beyond the Planck time to the moment of creation itself. The first 10^{-43}-second interval could be very interesting, involving physical processes never seen in the universe since that time. It could also be quite dull, little more than an extrapolation of what we already know. Until the theory is developed, we just will not know.

Once a quantum theory of gravity is available, a second question arises. Can we produce a truly unified theory of nature, one that includes gravity with the other fundamental forces? One problem with developing such a theory has to do with the enormous difference between the conceptual frameworks of general relativity and of quantum mechanics. In general relativity, gravity arises through a process that is essentially geometrical, while the grand unification demands that we regard all forces as being generated by the exchange of particles. We can discuss the classical problem of the orbit of the moon from the two points of view to illustrate this difference in viewpoint.

Using the concepts of general relativity, we would analyze the situation by saying that the moon circles the earth because of the effects that the earth's mass has on space. One easy way to picture this effect is to imagine empty space as a flat rubber sheet marked off in squares. The presence of the earth corresponds to a heavy weight placed on the sheet. In the region of the weight the sheet will be pulled down, creating a valley as shown in Figure 56. We say that the presence of the weight warps the surface of the sheet. If we represent the moon by a marble rolling on the sheet, then the motion of the moon in its orbit can be thought of as being analogous to the rolling of the marble around the side of the valley created by the earth.

The key point in this discussion is that nowhere does the concept of force or the concept of interaction enter the picture. The motion of the moon is explained totally in terms of geometry. The presence of the earth's mass alters the geometry of space, changing it from a flat plane to a valley. The orbit of the moon, then, is described in terms of the motion of another mass in that valley. This analogy captures the spirit of general relativity, although the actual calculations have to be done in a four-dimensional space time instead of on a two-dimensional rubber sheet.

If we were to approach the problem of the moon's orbit from the point of view of elementary particles, however, we would have to ask about the particles whose exchange generates the force between the earth and the moon. There is a precedent for this question in our discussion of electricity and magnetism in Chapter 5. There we saw that the electrical force between two charged particles was mediated by the exchange of a spin-1 massless particle called the photon. We know that one feature of the

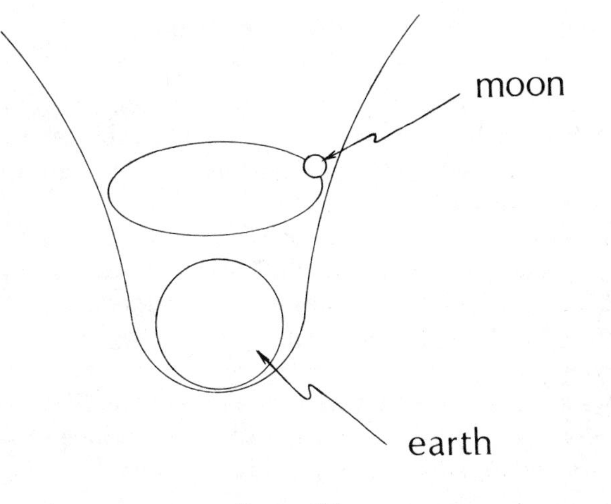

*Figure **56.***

electrical force is that like charges repel each other, while unlike charges attract. It turns out that this behavior is a general rule for forces associated with the exchange of spin-1 particles—like particles must repel.

With gravity, however, the behavior is different. Here every massive object attracts every other massive object—particles attract each other, antiparticles attract each other, and particles attract antiparticles. There is no repulsive gravitational force—no "antigravity." It can be shown that such a universal attraction can only arise through the exchange of spin-0 or spin-2 particles, but not through spin-1 exchange. This means that any quantum theory of gravity that we develop will have to involve the exchange of particles unlike the photons, gluons, and vector bosons we have encountered up to this point. In fact, it turns out that Einstein's theory of gravitation can only be reproduced by a theory that involves the exchange of a spin-2 particle, with no spin-0 particle involved at all. Since general relativity has stood up well to the experimental tests to which it has been subjected, we will take it as given that any quantum theory of gravitation will involve the exchange of a massless spin-2 particle. In Chapter 5 we included this particle, called the graviton, in our list of exchanged particles when we discussed the fundamental forces.

If we apply this sort of idea to the earth–moon problem, our notion of how the moon stays in orbit changes drastically from the one associated with Einstein's theory. The force between the earth and the moon is generated by a flood of virtual gravitons being exchanged between the two bodies. This force is completely dynamical in nature and has nothing whatsoever to do with geometry. It does, however, bear a strong resemblance to the other fundamental forces, at least insofar as it is generated by the exchange of a particle.

The development of a theory of gravitation based on the exchange of gravitons and the unification of this theory with the GUT is the task that this generation of theoretical physicists has undertaken. What the final form of the theory will be when (and if) they accomplish their goal remains to be seen, but there are some general features that we could expect the theory to possess. These general features, in turn, give us some idea of what to expect when we move beyond the Planck time.

Supersymmetry, Supergravity, and Superunification

There are really two parts to the task at hand. One is to develop a quantum theory of gravitation based on particle exchange; the other is to incorporate this theory in the grand unification. The situation is similar to what we encountered when we talked about unifying the strong force with the electroweak. First we had to have a theory of the strong interactions (quantum chromodynamics), and then we had to find a way to unify that theory with the Weinberg-Salam theory to produce the GUT. The technique being explored now is called supersymmetry, while the quantum theories of gravitation using supersymmetry are called supergravity theories.

Although there are conceptual difficulties involved in bringing about a unification of gravity with the other fundamental forces, we are not completely in the dark as to how to proceed. We have already seen that the use of the gauge principle led to the unification of the weak and electromagnetic forces into the electroweak interaction and to the unification of the electroweak and strong forces in the GUT. With this record of success, it is obvious that the first avenue to be explored should be the application of the gauge principle to gravitation.

Both the previous unifications share an important property. In each, we focused on some property that we labeled an internal symmetry and demanded that our theories of nature not change when we allowed this property to be defined arbitrarily by observers located at different points in space. For the electroweak unification, this property was the electrical charge (or, more precisely, the isotropic spin). When we talked about protons and neutrons, for example, we imagined that there was a little dial attached to each particle. If the dial pointed up, the particle was a proton; if it pointed down, the particle was a neutron. The principle of gauge symmetry said that our theory had to be unchanged if someone went around turning dials at random. It was this requirement, together with the process of spontaneous symmetry-breaking, that led to the Weinberg-Salam theory of the electroweak interaction.

When we wanted to unify the strong force with the electroweak, the relevant property was color. In this case the dial had three settings (for the three colors), and the theory had to remain unchanged if those dials were moved at random. In both of these unifications, the theoretical changes that would have occurred had the dials on the particles been reset were canceled by corresponding changes in the particles being exchanged to generate the force. The net result was a theory that was independent of the definitions made by different observers.

One way of thinking about the unification process, then, is to say that it relegates seemingly important differences between particles to being nothing more than arbitrary definitions. In a sense, the unification process strips away the surface features of the world and allows us to see what is truly important. By the time we have reached the grand unification, this process has led us to a world much simpler than our own. In this world there are two classes of particles: spin-½ particles, like the quarks and leptons, which we lump together under the generic term *fermions;* and spin-1 particles like the *X*-bosons and photons, which we shall refer to collectively as bosons. There are two kinds of forces operating between these classes of particles—the strong-electroweak and the gravitational.

Supersymmetry theories act in such a way as to erase the distinction between bosons and fermions. By analogy with the other unifications, you can imagine a dial attached to each elementary particle. If the dial points up, the particle is a fermion, and if the dial points down, the particle is a boson. Just as with the other unifications, we demand that our theory remain unchanged when these dials are reset at random, and just as in the other unifications, we find that we can make everything work out if certain types of particles are exchanged to produce the basic force.

In a scheme like this, we can think of the particle-plus-dial system as being a superparticle, and we can always imagine interactions in which bosons are transformed into fermions and vice versa in much the way that quarks were transformed into leptons by the actions of the *X*-bosons. In other words, although the terms *boson* and *fermion* are still meaningful, the interactions allowed by the supersymmetry theory allows us to change particles from one classification into the other. To a physicist, this means that the two classes of particles are really just different aspects of a single underlying fundamental entity.

So, the superunification can be expected to produce a universe that is the ultimate in simplicity. There will be only one kind of basic particle, the superparticle, and one kind of force acting in nature, the unified gravitational-strong-electroweak force. Thus, for the first 10^{-43} second of its existence, the universe was as simple as it could possibly be. Since that time, the freezings caused by the universal expansion have acted to produce more and more differentiation. Only the intense intellectual effort of generations of physicists has allowed us to see the true simplicity of it all.

The basic problem with the supersymmetry theories that are being investigated now is that they seem to predict the existence of many kinds of particles that have not been seen in the laboratory. On the one hand, this may not be too surprising, since one can always invoke the mechanism of spontaneous symmetry-breaking to say that the masses of these particles is very high—perhaps as much as 10^{19} GeV—so that they would never be seen. This is what happened to the *X*-particles that came in with the grand unification. On the other hand, the existence of all these particles does make for a complicated theory.

We can appreciate this point by thinking about the simplest version of supergravity. At low energies, gravity is a force mediated by the exchange of a massless spin-2 graviton, as we have explained and shown on the left in Figure 57. At higher energies, the theories predict that another massless particle, called the gravitino, will start to become important. Gravitinos have spin $^{3}/_{2}$ and would contribute to the gravitational interaction as shown on the right of Figure 57. (For technical reasons, gravitinos have to be exchanged in pairs and not singly.) Thus, when gravitating objects are far apart, as is the case with the earth and the moon, these theories produce the same prediction as normal general relativity. When objects are driven close together, as they would be in the violent collisions characteristic of the first 10^{-43} second,

graviton

gravitino

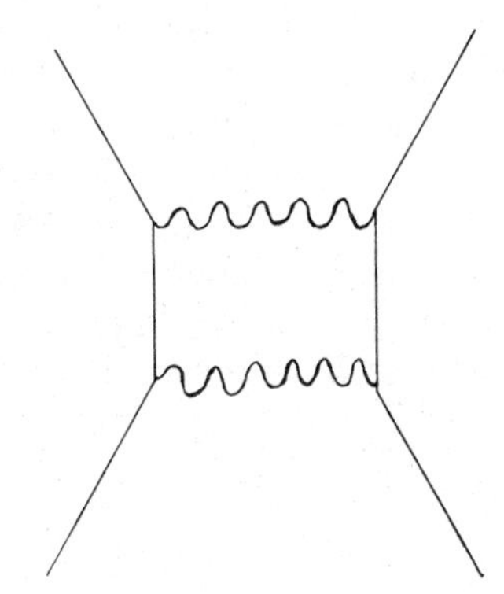

Figure 57.

gravitino exchange becomes important and modifies the gravitational interaction, eventually raising its strength until it is comparable to those associated with the other forces.

The general features of a superunification, then, would be the unification of the gravitational force with the other fundamental forces and the introduction of reactions in which fermions and bosons can be interchanged, resulting in a system in which there is only one kind of superparticle. Beyond these general features and the idea that the unification should occur at the Planck time, there are at present no firm theoretical guidelines in this area. On the other hand, it is clear that physicists are by no means puzzled about how to proceed in this last step toward the ultimate unified theory. It is just that the step is technically difficult and has not as yet been completed.

Ultimate Questions

This discussion of supersymmetry gives us a tantalizing insight into what physics might be like on the other side of the Planck barrier. But as interesting as a fully unified theory might be, the prospect that such a theory might soon be developed leads us naturally to the ultimate question. Once we have traced the universe back to the moment of creation itself, what is to keep us from taking the next step and asking about what happened before the Big Bang—to ask, in effect, why the universe exists at all?

That is a question that scientists feel a little uncomfortable asking. Although anyone who thinks about cosmology and the early universe must eventually think about that question and although anyone involved in bringing the results of science to the general public will eventually be confronted with it, the question is usually evaded in one way or another. In fact, we can identify two general ways of dealing with the ultimate origin of the universe that have evolved—one derived from the philosophy of science and the other from general relativity.

All scientific laws are based on observation and experiment, and consequently, no scientific law is really valid outside of the domain in which it has been tested and verified. It is possible to argue, therefore, that the question about the origin of the universe simply cannot be answered within the scientific method. We can discover laws by making measurements and observations, and it is probably valid to extrapolate these laws, as we have done, to the early universe. But we have no experience whatsoever with a universe that does not contain mass, and it is therefore improper to try to extend our present knowledge to this new area. The question of the origin of the universe, according to this argument, cannot be answered by the methods of science and must therefore be left both unasked and unanswered.

The argument from general relativity is somewhat more technical. We know that in the framework of that theory the four dimensional space-time in which we live is the result of the presence of mass. We saw an example of this when we talked about the orbit of the moon. In this way of looking at things, time began when mass was created, and asking about what happened in times previous to that is simply meaningless. As one set of authors suggested, asking what came before the Big Bang is like asking what is north of the North Pole.

Both of these positions are perfectly defensible; indeed, I have often used the first one myself when pressed by my students. Nevertheless, it does seem to be a bit disappointing to have come so far in our study of creation only to be brought up short on the verge of solving the most interesting problem of all. Furthermore, the refusal to examine this, the ultimate question of creation, seems to me to fly in the face of the entire trend of the grand unification scheme.

We have seen that the central theme of the unification process is the requirement that the laws of physics should be independent of arbitrary definitions made by outside observers. We could argue that this requirement is a logical extension of the principle of relativity, which holds that the laws of physics must not depend on the state of motion of the person observing events. Both of these great advances in twentieth-century science seem to point in the same direction: in an uncertain world, the one absolute, the one firm point of reference, is the body of knowledge we call the laws of nature. These laws do not depend on the state of motion or the state of mind of the observer, and hence everyone who looks into nature will find the same laws that we have found. It is not too great a leap, then, to suggest that we use this firm bedrock to anchor ourselves when we try to push human knowledge where it has never gone before. Although the creation of the universe may involve a process we have never seen (and never can see), let us assume that the laws of nature that we have discovered can be used to think about it.

If we ask what could have occurred before the Big Bang, there are only two possibilities. One is that the universe proceeds in cycles, where phases like our present expansion alternate with periods of contraction. We will discuss this possibility in some detail in Chapter 14, but for the moment, we simply note that if this happens, the question of ultimate beginnings becomes pointless. The universe always was and always will be, with one cycle following another forever. This view was, of course, widely held by ancient philosophers.

The other possibility is that the Big Bang was a unique event. In this case, the best guess as to what preceded it is that prior to the Big Bang there was nothing—a vacuum. But one thing that we have learned from quantum mechanics is that there is no such thing as empty space with nothing whatsoever in it. The uncertainty principle (see Chapter 5) guarantees that even in the best vacuum, virtual pairs of particles and antiparticles will be created and annihilated continuously. This means that the vacuum is a system like any other, and the question of why the universe exists at all can be cast in a physically meaningful way. The empty universe (the vacuum) has an energy that we could, in principle, calculate. Similarly, the universe at the moment of creation (that is, a universe with mass in it) also has an energy. If the latter has a lower energy than the former, then the vacuum will be unstable in a very real sense. Our understanding of the laws of nature is that every system will proceed toward a state of lowest possible energy, and the vacuum is no exception to that rule. In the words of Frank Wilczek of the University of California at Santa Barbara,

"Perhaps the reason that there is something instead of nothing is that nothing is unstable."

The idea that we can describe creation itself with the same laws of nature that we discover in the here and now has been taken seriously by enough people so that we can provide some insight into scientific thinking on this subject. Some of the most interesting work on this topic has been done by two theoretical physicists at the Free University of Brussels in Belgium. Francois Englert and R. Brout use a delicate interplay between matter and cosmological expansion to describe creation. In essence, they treat creation with the same concepts that would be used to describe the balancing of a pencil on its point. We know that it is possible to balance the pencil that way, but we also know that if there is any slight deviation from verticality, the force of gravity will act to pull the pencil further over. The net effect of even the smallest deviation is that the pencil will fall down.

We can think of this instability as arising from a sort of vicious circle. First, the pencil, for one reason or another, deviates a little from the vertical. This causes gravity to exert a force that tends to rotate the pencil around its point, which in turn causes the pencil to lean further from the vertical, which increases the pencil's susceptibility to gravity, which causes still more of a deviation, and so on. This sort of reinforcing effect is typical of unstable systems.

Brout and Englert argue that a similar situation exists in a massless universe. If we imagine empty space to be a rubber sheet with a grid marked out on it (as we did for our description of general relativity), then we can also imagine that there will be small perturbations in the sheet. If nothing else, the quantum mechanical process of pair creation and annihilation would produce such an effect. Brout and Englert argue that if at any point a small expansion of space is started (think of this as a microscopic stretching of the rubber), the effect would be the same as a small tilt of the balanced pencil. In their theory, the presence of the expansion would lead to the creation of a small amount of matter. This creation process would be by ordinary quantum mechanical means. The matter would then cause more expansion, which would create more matter, which would cause more expansion, and so on. Like the falling pencil, the instability of the universe would run away until enough matter has been created for the universe to appear as we would expect it to appear at the Planck time. In this scheme of things, there was no Big Bang but only a slow buildup of matter until the density was high enough to put the universe into the false vacuum (see Chapter 11). From this point on, the history of the universe would be identical to that of any inflationary system.

The intriguing thing about this scheme is that the question "What happened before the Big Bang?" is answered by the assertion "There was no Big Bang." Time does not begin at a particular point but stretches back to minus infinity. Until the end of what Brout and Englert call the creation epoch, the process of matter creation from the basic instability goes on. But before the Planck time, enough matter has accumulated to change the character of the system and the rest of the expansion goes on as we have described it in previous chapters. The turnover point in the early universe is analogous to the point at which the falling pencil hits the table. The new situation produces a new kind of behavior.

We can get an idea of how this sort of model compares to the Big Bang models we have been discussing by making a graph of the density of the universe as a function of time, as shown in Figure 58. The Big Bang is described by an expansion starting from an infinitely dense point and proceeding smoothly from that point on. This is shown in Figure 58 by the falling line, indicating a uniform drop in density as the expansion goes on. (The density must fall in this case, since the same amount of mass now has to fill a larger volume.) By contrast, the inflationary scenario discussed in Chapter 11, although it also starts from a single infinitely dense point, goes through a period of rapid expansion with constant density. This is what we called the inflationary era and is represented by the horizontal line on the graph. Eventually, the universe tunnels out of the false vacuum, great numbers of particles are created, and the inflationary scenario becomes identical to the standard Big Bang scenario.

Both of these versions of the creation, however, start with a singularity—a point at which matter is in an infinitely dense state. This may make sense mathematically, but it is hard to see how we could apply any laws of physics to such a situation. The unstable-vacuum scenario we have just outlined, however, would have a gradually increasing density until it reached the false-vacuum state, at which point it would proceed just like the ordinary inflationary scenario.

Other theorists have proposed other ways of approaching the problem of creation while still avoiding the difficulties associated

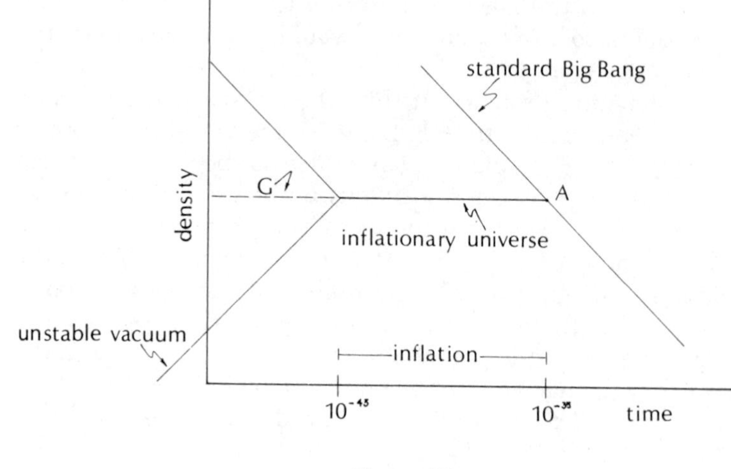

Figure 58.

with the singular point. Richard Gott of Princeton University, for example, has worked out a model in which the universe is created with a density appropriate to the false vacuum, rather than with either zero or infinite density. Such a model would be represented by the horizontal line labeled *G* in Figure 58. You could think of this as a model in which the universe tunnels into the false vacuum state only to tunnel out again later.

All of these views of creation have one thing in common: they envision the universe coming into being as a quantum fluctuation, a random event; once the event has occurred, the laws of physics take over and our present universe develops. The laws of physics even describe the fluctuation process but give no overriding reason as to why the universe must have come into being at all. It may be, then, in the words of physicist Edward Tryon, that "our universe is simply one of those things that happens from time to time."

Chapter

14

The Fate of the Universe

When You and I beyond the Veil are past
Oh, but the long, long time the earth shall last.
RUBAIYAT OF OMAR KHAYAM

"It ain't over 'til it's over."
YOGI BERRA

No discussion of cosmology would be complete if we restricted our attention solely to the question of the origins of the universe. Having traced our roots beyond the Planck time and seen where the frontiers of knowledge are in that direction, it is only natural to turn and look the other way, into the future. This changes the focus of our inquiry from where we came from to where we are going.

The primary barriers to gaining a clear answer to the first of these questions have been theoretical rather than experimental. We have had to develop and test theories capable of describing interactions at much high energies than any considered before. With the GUT, an important milestone on the road to ultimate understanding was reached, and there are high hopes that the idea of supersymmetry will lead us to a truly unified theory of elementary interactions. Some experimental tests of the unified theories have been carried out, and others are under way, but all of these tests involve laboratory processes. We had no need, in applying the GUT concepts, to get new observational data about the universe itself. The theories were invoked to explain facts already known; they did not seek new observations to guide their development.

When we look to the future the situation is reversed. General relativity and some basic results from quantum mechanics and thermodynamics are quite enough to describe the universe as far into the future as it is possible for us to see. What we do not have at present is a sufficiently accurate measure of the density of matter in the universe to be able to say with certainty which of two possible futures is in store for us. Therefore, progress in predicting the fate of the universe is dependent on our ability to improve our observational, rather than theoretical, skills.

Although the earth and the sun do not constitute a major part of our galaxy and although the Milky Way is only one galaxy among billions, we have nonetheless a certain interest in their futures. The sun is now about 5 billion years old. In another 5 billion years it will have used up all of its hydrogen fuel. It will become what astronomers call a red giant, and the earth will be swallowed up. Except for very small and very long-lived stars, 10 billion years is a typical lifetime for a star. Over this time period, the star takes primordial helium and hydrogen, processes it in its nuclear furnace, and dies. Depending on the mass of the star, the end product can be a white dwarf (this will be the fate of the sun), a neutron star, or a black hole. But whatever the end product of stellar evolution, each time a star goes through the life cycle, it uses up a certain amount of raw material. In 40 or 50 billion years, we expect that star formation will probably have slowed down considerably from what we see now. In our own galaxy, astronomers will see the stars go out slowly—first the bright ones (which burn up their fuel most profligately and therefore die soonest) and then more sedate stars like the sun. The period of star formation, which started with the appearance of galaxies 500,000 years after the Big Bang, will end.

The Universal Expansion

The next step in our survey of the future is to move away from a relatively parochial concern with our own immediate neighborhood and ask about the fate of the universe as a whole. As we saw in Chapter 1, the most striking feature of the universe taken as a whole is expansion, with every galaxy receding from every other galaxy. It is reasonable to ask whether this expansion will continue forever or whether it will someday stop.

The best way to think about this question is to imagine a ball thrown upward from the surface of the earth. We know that eventually the ball will slow down, stop, and reverse its direction because of the gravitational attraction of the earth. However, if the ball were thrown fast enough (more than 7 miles per second), we know that it would not fall back but would sail off into space instead. The general question of whether the ball will fall back to the ground depends on two things: how fast it is moving and how hard the earth is pulling on it. The force of gravity exerted by the earth, in turn, depends on how much matter the earth contains.

We can think of the present expansion of the universe in the same way. A given galaxy is now receding from us at a particular velocity, which we can measure. Whether it will ever stop and start falling back toward us depends on how much of a gravitational attraction the rest of the universe exerts on it. If there is enough matter to exert a strong enough force, we can expect that the outward-rushing universe will someday start to contract. If there is not enough matter in the universe, then the expansion may slow down a bit, but it will never stop. The first question we must ask in charting the future of the universe, therefore, is whether the universe is open, closed, or flat.

This particular way of looking at the question, while it is easy to visualize, is slightly inaccurate from the point of view of general relativity. In Chapter 13 we saw that the gravitational force is thought of in terms of the bending of space by the presence of matter. The analogy of the distortion of a rubber sheet is an easy way to picture this way of looking at things. The more mass, of course, the more distortion there is.

In terms of this way of treating gravity, a closed universe would be one in which the amount of matter is so great that the rubber sheet is bent around on itself, so that no body can ever be an infinite distance from any other. Similar definitions apply to the terms *flat* and *open.*

It might seem strange that we do not know enough about the structure of the universe to answer this fundamental question. After all, our telescopes are capable of detecting galaxies over 10 billion light-years away. Why can't we just add up all the mass that we see and get an answer?

If we simply add up the estimated masses for all of the galaxies and stars that we can see with our telescopes, we would have only a small percent of the amount of matter needed to close the universe. It would appear that we have to conclude that the universe is open.

But this statement is premature, for it depends on the assumption that we can see all the matter that there is. We know this is not always true. For example, someone looking at the solar system from another star would see the sun because it is luminous, but he would probably not see any of the other massive bodies we know are here—planets, asteroids, comets, and so on. In the case of the solar system, this probably is not too important, because all of these bodies add up to only a tiny fraction of the mass of the sun. But in looking at galaxies, the situation may be different. It may well be that a large percentage of the mass in a galaxy is not visible to someone looking at it through a telescope. If the matter we see does not close the universe, then perhaps the matter we do not see will do the job.

There are several pieces of evidence that there might be a great deal of as yet undetected matter. One bit comes from studies of galaxies themselves, and the other from studies of clusters of galaxies. Galaxies like the Milky Way and our nearest neighbor, Andromeda, are shaped like spirals and rotate around their centers. The sun, for example, makes a grand circuit around the Milky Way every 200 million years or so. This rotation is generally very complex, with different parts of the galaxy going around at different speeds. There is, however, one aspect of galactic rotation that we can be sure we understand, and that is the way matter should behave when it is at the outermost edges of the system. Just as the outer planets, such as Jupiter and Saturn, move more slowly in their orbits than Earth, so too should matter in the outer fringes of a galaxy exhibit a gradual slowing down as the distance from the galaxy increases. We say that the rotation should become Keplerian (after Johannes Kepler, the man who first described the behavior of the planets in the solar system).

When we speak of the "outer fringes" of a galaxy in this context, we are talking about the tenuous material that lies beyond the farthest star. This material, mostly hydrogen atoms, does not emit light, so it is not visible to the naked eye or an ordinary telescope. It does, however, emit radio waves, which allow us to detect its presence and determine its speed of rotation. These radio maps of the galaxies reveal a rather startling fact. In most galaxies, the wispy hydrogen gas is rotating just as fast as the stars and shows no sign of the expected Keplerian behavior. It is as if we measured the solar system and found that Jupiter was moving just as fast as Earth.

The only way this puzzling fact can be explained is to assume that the hydrogen gas we see is being affected by a large collection of matter in the galaxy; matter that remains undetected except for its effect on the hydrogen. Every galaxy, then, is surrounded with an extensive halo of invisible matter, a halo that could very well contain more material than the stars themselves. Most estimates place the mass of the halos between two and ten times the visible mass of the galaxy.

What the galactic halos teach us is that we should not be too hasty in declaring the universe to be open. Even though we can see less than a few percent of the matter required to reverse the universal expansion, we now know that there is surely a great deal of matter in the universe that we cannot see. If we assume that most galaxies have halos, then we should multiply the amount of visible matter by a number between 2 and 10 to get an idea of how much mass is really out there. Of course, doing so does not get us to the critical amount of mass by a long shot, but this episode does make us wonder whether there are not other unseen masses waiting to be discovered.

Another candidate for unseen matter arises from astronomers' studies of galactic clusters, which are large concentrations of matter containing thousands of galaxies like the Milky Way. Observations of these clusters show that they are much too heavily populated to be simply a random collection, but that the constituent galaxies are moving far too rapidly to allow the cluster to stay together for much more than 100 million years, a very short time on the cosmological scale. The only way to explain the appearance of these clusters (as well as some of their detailed structure) is to assume that there is more matter in the cluster than can be seen. Typically, a cluster will have ten to forty times more matter in it than one would guess if one just looked at luminous material. Most of this will be halos, of course, but even counting halos there is extra unseen mass in galactic clusters, a fact that should make us all the more reluctant to conclude precipitously that the universe is open.

All of this discussion of unseen matter is concerned with determining the density of matter in the universe so that, knowing the present rate of expansion, we can decide whether the universe is open or closed. It should be clear, however, that knowing the matter density and rate of expansion at *any* time would decide this question just as well. By considering the universe during the period of nucleosynthesis three minutes after the Big Bang, we can come up with a reasonable, albeit indirect, statement on this subject. The rate of expansion at that time is easy to calculate, and the density of protons and neutrons can be inferred from the amount of deuterium and helium that was produced. (You will recall that it was deuterium production that was the bottleneck in the creation of nuclei at that time). Subject to some qualification, such as the assumption that the deuterium abundance on earth is typical of the deuterium abundance everywhere, the result that comes from this sort of calculation is unambiguous. If we count only matter in the form of baryons, the universe is open. It follows from this that if the universe is to be closed, the extra mass must be in some form other than ordinary protons and neutrons.

All of the evidence we have, then, seems to indicate that the mass of the universe is very close to the critical value needed for a barely closed (or open) system. This state of affairs is what we earlier called the flatness problem. Furthermore, as the discussion of hidden mass shows, the true value of the mass density is likely to be much closer to critical—perhaps within a factor of 2.

If we want to talk about the future of the universe, however, this statement is not good enough. We need to know whether the expansion will reverse itself or not, and given the present state of our observational knowledge, we cannot make this distinction. Consequently, we will have to think about both alternatives and see what the future holds in store for each.

The Closed Future

Let us begin by assuming that for some reason or other nature has chosen to hide 95 percent of the mass of the universe in places very hard to spy out and that the universe is actually closed. In this case we are in for a spectacular future. For another 40 or 50 billion years, the universe will continue to expand, but ever more slowly. Then, like the ball falling back to earth, the expansion will reverse and a great contraction will begin. Instead of a universe in which light from distant galaxies is shifted toward the red (indicating that the source is receding from us), we will find a universe in which light is blue-shifted.

When the universe is again its present size 80 or 100 billion years from now, the earth and the sun will be long dead. If there are any human beings around at the time, they will undoubtedly be living in man-made environments circling stars whose births are still unimaginably far in the future. As we have argued, the galaxies will be decidely less luminous than they are now, with large populations of white dwarfs, neutron stars, and other very faint objects. As the contraction progresses, the night sky will slowly begin to brighten as the cosmic background radiation begins to shift toward the visible part of the spectrum, and eventually the sky will blaze with light, day and night. By this time, the universe will have contracted to less than one-thousandth its present size. Atoms and molecules in interstellar space will dissociate into their constituent nuclei and electrons, and eventually the stars and planets themselves will dissolve into a universal sea of hot material. From this point on, the stages of the Big Bang that we described earlier will simply replay themselves in reverse—nuclei dissociating into protons and neutrons, protons and neutrons dissociating into quarks—until we are back to the original state that started the Big Bang.

This scenario leads inevitably to the most fascinating question of all: Will the universal contraction (which cosmologists half-jokingly call the Big Crunch) be followed by another expansion (the Big Bounce)? In other words, will the universe arise phoenix-like from its ashes and repeat the entire cycle? The picture of a universe that is reborn every hundred billion years is very attractive to some people. The main advantage of an eternally oscillating universe is that the questions of why it all started and where it all came from simply do not have to be asked. The universe always was and always will be. A hundred billion years from now the universe will again consist of a large collection of separating galaxies.

It is a fascinating thought, but before we go too far into speculation, I should warn you that there are some serious problems with the oscillating-universe picture. For one thing, unless some of the basic laws of physics change during some part of the cycle, the average disorder of the universe would have to increase during each bounce, so that eventually the system would have to run down. And, of course, the whole idea of oscillations depends on the presence of enough mass to reverse the expansion and initiates the Big Crunch. As we have seen, our present data seems to favor a quite different type of future.

The Open Future

If the universe is open or flat, then no reversal of the expansion will occur and the future will be truly infinite. On the near term, the sun will burn out and star formation will slow down just as we have described. Small, slow-burning stars will just be going out when the age of the universe reaches 10^{14} years—about a thousand times the present age. On longer time scales, other kinds of dissipation begin to become important. For example, every 10^{15} years or so, stars with planetary systems can be expected to come close enough to each other for the planets to be torn loose. Stars will evaporate from the outer regions of the galaxy in a time scale of 10^{19} years, while the densely packed stars in the galactic center will collapse together into a large black hole. When the universe is a billion times older than it is now, we will see an ever thinner sea of background radiation in which an occasional black hole is embedded. Scattered around among these landmarks in nothingness will be the solid remains of the evaporated stars and such dust and miscellaneous debris as has escaped capture up to this point. The universe will keep this aspect until its age reaches 10^{31} years.

At this point we have to ask about the fate of the remnant of solid matter in the universe. According to the GUT, the protons and neutrons that make up all matter are unstable and will disintegrate with a lifetime around 10^{32} years. If matter is indeed unstable on this very long time scale, than at this point the remaining solid matter will disappear as its constituents decay. The net long-term effect of this decay will be to eliminate solid matter from the universe, producing some extra radiation and widely separated electrons and positrons in the process.

The universe will go on expanding and cooling off. Occasionally, some of the miscellaneous particles will fall into a black hole, producing X rays as they do so. A hypothetical astronomer observing the universe would be getting very bored, because this state of affairs would persist until 10^{65} years had passed.

We are now getting into such long time scales that it becomes almost impossible to form mental pictures to go with all those zeros. One useful visualization is to imagine watching the universe on a film. Suppose we started by letting the film show 10 billion years of history each minute. It would then take somewhat less than 2 minutes to see everything up to our present era, some 15 billion years after the Big Bang. We would let the film run at this speed for 10 minutes, a process that would take us to an age of 100 billion years. If the universe is closed, the Big Crunch would be almost over by this time. After watching for 10 minutes, we would increase the film speed by a factor of 10, so that each minute of viewing corresponded to 100 billion years, and 10 minutes would take us to an age of a trillion (10^{12}) years. It would take three such changes of film speed to cover the burning out of small stars, another to see the planets torn loose, and four changes beyond that to see the galaxies dissolve. It would take 13 more increases in film speed (21 in all) to get us to the point where protons are expected to decay, and we would then sit and watch no fewer than 34 changes in speed before the next interesting event.

We normally think of black holes as bodies so massive that nothing can ever escape their gravitational pull, not even light. Yet on long time scales it turns out that this is not quite accurate. Black holes will give off thermal radiation when the temperature of the background radiation is low enough. In a sense, the black hole is like the ember of a fire, giving off heat to its surroundings—heat that in the case of the ember you could feel with your hand. When the film was running at the rate of 10^{65} years per minute, a black hole the size of the earth would appear to start radiating energy, getting brighter and brighter as it does so. In 1 minute of film time, the black hole would brighten the sky and then disappear, its only monument an addition to the expanding sea of radiation. As the film ran on, speeding up every 10 minutes, larger and larger black holes would undergo the same process and evaporate themselves away, and for the next forty changes in film speed, that is what we would see—an expanding universe with occasional fireworks as various black holes die. This process would go on until all black holes are gone, and by the time it was over, the film would have been running at the speed of 10^{105} years per minute. We would have been watching the film for a little less than 18 hours.

At this point we would have reached the end of our story, because there would be nothing left in the universe to produce any change. This means that all possible futures in the open universe end up in the same way. At some distant time in the future, the universe will be a cold, thin, expanding sea of radiation, with a few forlorn particles to break the monotony. Perhaps it was this gloomy prospect that caused Steven Weinberg to remark, "The more the universe seems comprehensible, the more it also seems pointless."

Epilogue

Now that we've settled the beginning and end of the universe, all we have to do is make it through to Friday.
DON MOSER

Defining the Frontier

After reviewing the tremendous strides that have been made in our understanding of the early universe, it is natural to ask what is likely to happen now and in the near future. At the moment, we can picture the frontier of knowledge as a rather fuzzy line through the general area of the grand unified theories. An optimist might argue that the GUT falls well within the boundary of established science, with the remaining problems more in the nature of a mopping-up operation than a major battle. A pessimist might point to the fact that there is as yet no firm experimental evidence for the GUT, nor is there agreement on such important matters as the nature of the true underlying symmetry. In terms of our Big Bang timetable, the optimist would argue that we have pushed all the way back to the Planck time at 10^{-43} second, while the pessimist would opt for something closer to 10^{-35} second. Both, however would have to agree that the boundary has been moved back a great deal from where it was a decade ago. The fuzziness as to the exact location of the frontier of knowledge is not particularly important. Indeed, it is what we would expect of a discipline that is undergoing rapid growth.

We have already outlined the most likely next step in the process of unification. The theoretical problem is twofold: first, to develop a quantum mechanical description of gravity and, second, to unify that theory with the GUT. This is a subject of intense research effort right now. If these efforts are successful, we will be able to push our knowledge of the early universe back to the moment of creation itself. What we find when we get there and whether we can describe the creation process itself are subjects of speculation at the moment. Our present knowledge does not permit us to see clearly what the answers to these questions will be.

These are the immediate unsolved theoretical questions. There is another set, more fundamental, that can be asked. For example, we know that the unification masses for the electroweak and grand unifications respectively, are 100 GeV and 10^{15} GeV. Why should they be so completely different? Is it really true that after we finish building the present round of accelerators, capable of probing interactions in the 100-GeV range, we will be faced with a great desert in which nothing interesting happens until the impossibly high energy of 10^{15} GeV? If this should turn out to be the case, it would be an event unprecedented in the history of physics. After all, the difference between the energy required to break up an atom and the energy required to break up a nucleus is only five or six orders of magnitude (the former being a few electron volts and the latter a few million electron volts). To go from the energy required to break up nuclei to that required to produce particles is only three orders of magnitude (from million electron volts to GeV), and two orders of magnitude beyond that we find the electroweak unification at 100 GeV. A gap of thirteen orders of magnitude before the grand unification would be very difficult to understand. Theorists call this the problem of scale. It is unlikely to be solved soon.

The Triumph of Reductionism

There is an important aspect of the advances in physics discussed in this book that we have not yet mentioned. These advances represent the final working-out of an old scientific and philosophical goal. Western science has been largely based on the idea that the way to understanding anything in the physical world is to break it down into its constituent parts. Thus, materials are understood as being composed of atoms; atoms, of particles; particles, of quarks; and so on. This is a way of looking at things that philosophers call reductionism. The basic assumption of reductionism is that the underlying reality will be simple and beautiful and that the apparent complexity of the world is the result of complex relationships between simple objects. Thus, a brick is a simple thing, but bricks can be put together to make all sorts of intricate structures.

During the 1960s and 1970s, when the complexity of the particle world was being made manifest in one experiment after another, some physicists broke faith with the reductionist philosophy and began to look outside of the Western tradition for guidance. In his book *The Tao of Physics,* for example, Fritjhof Capra argued that the philosophy of reductionism had failed and that it was time to take a more holistic, mystical view of nature.

But we now see that this point of view, generated in what turned out to be a temporary era of growing complexity, has not stood the test of time. We have seen that the introduction of the idea of quarks has reduced the complexity of the elementary particles and that the development of the idea of gauge symmetry has reduced the apparent complexity of the fundamental forces. We stand, in fact, on the verge of the ultimate triumph of the reductionist idea. Within a few years, we may very well realize that if we probe deeply enough, we will find a universe that is the ultimate in simplicity and beauty. All of the apparent complexity we see will be understood in terms of an underlying system in which particles of one type interact with each other through one kind of force. If this sort of theory is actually developed, it will be the culmination of a philosophical quest that began over two millennia ago in the Greek Ionian colonies.

Another important tradition in Western science has been the separation of the observer from what is observed. In classical physics this idea took the form of demanding that the act of observation have no effect on the system being observed. As we have seen, the uncertainty principle shows that this rigid separation of subject and object cannot be carried over into the subatomic world. In that world, the act of observation necessarily implies an interaction of one particle with another and that interaction must disturb both particles. Effects such as virtual particle exchange and quantum tunneling are seen to be measurable and verifiable consequences of the uncertainty principle.

But there is a larger sense in which the development of gauge theories reinforces the idea behind the separation of subject and object that is so much a part of our everyday outlook on life. It seems to me that the essential point underlying this separation is that there is actually a physical world that can be known and that this knowledge must be independent of the person who knows it. This cannot be established at the level of experimental events because of the uncertainty principle, but the gauge theories seem to say that it can be established at a much deeper and more profound level.

The gauge theories imply that the correct descriptions of nature are those that do not depend on arbitrary definitions made by different observers. In fact, the theories require that there be no measurement which could be made in the laboratory which would be different if every observer in the world made an arbitrary change in his definitions of electric charge, color charge, or even elementary particles. The gauge principle, therefore, tells us that the only theories which can correctly describe the natural world are those which are completely independent of the state of mind of different observers. In this, they echo the earlier successes of the theory of relativity, which requires that there be no law of nature that depends on the state of motion of individual observers. The net effect of these two theories is to open a new and deeper gap than ever existed between the world being observed and the one doing the observing. In effect, they tell us that the correct description of the world is that description in which the observer is irrelevant. Any other description would violate the gauge principle and would produce an incorrect theory.

The 1970s, then, can be thought of as the period in which the great traditions of Western scientific thought, seemingly imperiled by the advances of twentieth-century science, have been thoroughly vindicated. Presumably, it will take a while for this realization to percolate away from a small group of theoretical physicists and become incorporated into our general world view.

What About God?

When I talk to my friends about the fact that the frontiers of knowledge are being pushed relentlessly back toward the moment of creation, I am often asked about the religious implications of the new physics. That there are such implications is obvious, particularly in the speculations about how the universe came into existence in the first place. Physicists normally feel very uncomfortable with this sort of question, since it cannot be answered by the normal methods of our science. For what it is worth, I will give my own personal views on the subject here, with the caveat that these views may not be shared by other scientists.

It seems to me that the unease people feel when they think about the sort of scientific advance implicit in the new physics arises from a notion that applying the techniques of science to the creation of the universe is somehow encroaching on terrain that has been staked out by religion. I imagine that this is how our ancestors felt in the nineteenth century when Darwin finally bit the bullet and applied the laws of evolutionary biology to human beings. Yet, in retrospect, we can see that the fact that human beings evolved from lower life forms does not damage the central tenets of anyone's religious beliefs. To take Christianity as an example, evolution is simply irrelevant to the doctrine of salvation through faith or any other important teaching.

I suspect that if we are successful in understanding the mysteries of the moment of creation, our descendants will have much the same attitude toward that fact as we have toward evolution. The reason for this expectation has to do with the nature of the scientific process. No matter how deeply we probe into any scientific subject, we will always find something unexplained and undefined. Medieval philosophers took the earth as given and attributed its existence to the special creative work of God. In the nineteenth century, it was realized that the existence of the solar system followed naturally from the law of gravitation and the existence of the galaxy, and in this century we have discovered that the existence of the galaxy is a natural consequence of the Big Bang. In all of these cases, the frontiers of knowledge have been pushed back by the discovery of new laws of nature. At each step, however, there was a point at which one could say, "Our scientific knowledge has brought us this far; beyond this point we may, if we wish, postulate a special creation."

It now appears that our new discoveries of the laws that govern the nature of elementary particles may allow us to push the frontiers back to the very creation of the universe itself. This does not, however, alter the fact that there is a frontier. All it does is transfer our attention from the material form of the universe to the laws that govern its behavior. I can hear a twenty-first century philosopher saying, "Very well, we agree that the universe exists because of the laws of physics. But who created those laws?" And even if, as some physicists have suggested, the laws of physics we discover are the only laws that are logically consistent with each other (and therefore the only laws that could exist), our philosopher would ask, "Who made the laws of logic?"

My message, then, to those who feel that science is overstepping its bounds when it probes the early universe is simple: don't worry. No matter how far the boundaries are pushed back, there will always be room both for religious faith and a religious interpretation of the physical world.

For myself, I feel much more comfortable with the concept of a God who is clever enough to devise the laws of physics that make the existence of our marvelous universe inevitable than I do with the old-fashioned God who had to make it all, laboriously, piece by piece.

Glossary

Annihilation The process that occurs when a particle and antiparticle meet and their masses are converted into energy.

Antimatter A form of matter that, at the particle level, consists of a particle whose mass is equal to that of a normal particle but which carries opposite electrical (and other) charges (*see* Chapter 3).

Antiproton A particle of the same mass and spin as the proton, but with a negative electrical charge.

Asymptotic freedom The property that allows quarks to interact freely while they are bound within particles.

Baryon Any particle whose decay products include a proton.

Boson An elementary particle having integer spin (*ie.*, spin 0, 1, 2 . . .)

Charge conjugation The mental operation in which we imagine replacing every particle with its antiparticle.

Color An internal symmetry that exists in quarks. It is analogous to (but not as easily visualized as) electrical charge.

Confinement The principle that leads to the conclusion that quarks must remain bound in particles and cannot be seen as free entities.

Conservation law A law that states that a particular quantity does not change (is conserved) in a given physical process.

Coupling constant The number used to specify the strength of the interaction between two particles.

CPT Theorem The theorem that states that every law of nature must be the same if we replace particles by antiparticles, observe in a mirror, and run the film backward.

Deuteron A nucleus made up of one proton and one neutron.

Deuterium The atom that has the deuteron as a nucleus.

Doppler effect The effect by which the motion of the source of a wave shifts the frequency of that wave (*see* Chapter 1).

False vacuum A state in which the universe might have found itself soon after the Big Bang and from which it tunneled out.

Fermion Any particle with half-integer spin (*i.e.*, spin $1/2$, $3/2$, $5/2$, etc.)

Flatness problem The problem associated with the fact that there is almost exactly enough matter in the universe to close it.

Gauge symmetry A symmetry that holds when no measurable quantity depends on definitions made by different observers.

Gauge theory Any theory that has gauge symmetry.

Generation A term referring to groupings in which the quarks or leptons are arranged in three pairs, each pair being a generation.

GeV Giga electron volt; a unit of energy corresponding to 10^9 times the energy needed to move one electron through one volt. The proton mass is a little less than one GeV.

Global symmetries A symmetry in which no measurable quantity depends on the definition of a particular quantity but in which all observers must use the same definition.

Gluon The massless spin-1 particle whose exchange generates the strong interaction between quarks.

Grand Unified Theory The theory in which the strong, weak, and electromagnetic forces are seen to be identical.

Gravitino A massive spin-$3/2$ particle whose exchange contributes to the force of gravity in some supersymmetry theories.

Graviton A massless spin-2 particle whose exchange is thought to produce the gravitational force.

GUT Acronym for Grand Unified Theory. Usually pronounced "gut."

GUT mass/energy The energy at which the strong, weak, and electromagnetic forces are unified. Usually taken to be 10^{15} GeV.

Hadron Any particle that participates in the strong interaction.

Horizon problem The problem associated with the fact that radiation coming from different directions in the universe is isotropic, even though the regions where it is emitted could not have communicated with each other.

Hubble age The approximate age of the universe obtained by extrapolating the observed expansion backward in time. The accepted value of the Hubble age is roughly 15 billion years.

Inflation The process by which the universe expanded very rapidly about 10^{-35} second after the Big Bang.

Internal symmetry Properties of elementary particles such as electrical charge, strangeness, charm, color, etc., that characterize the structure and behavior of the particle.

Ion An atom from which electrons have been stripped.

Ionization The process whereby electrons are torn off previously neutral atoms.

Isotropic A quantity is isotropic if it is the same when viewed from any direction. The microwave radiation is approximately isotropic (*See* Chapter 1).

Lambda A spin-½, strangeness −1 baryon. This was the first strange particle discovered.

Lepton A particle that does not participate in the strong interactions. The leptons are six in number, the electron, muon, tau, and the neutrino associated with each of these particles.

Light-year The distance that electromagnetic radiation can travel in a vacuum in one year, approximately 6 trillion miles (9 trillion kilometers).

Local symmetry A symmetry in which no measurable quantity depends on the definition of a particular quantity and in which each observer may make his definition independently of any other.

Magnetic monopole A very massive particle formed 10^{-35} second after the Big Bang and carrying an isolated north or south magnetic pole.

Meson Any particle whose decay products do not include a proton.

Microwave Electromagnetic waves similar in all respects to visible light except that they have a longer wavelength.

Mu meson (Muon) A lepton similar to the electron in every way except that it is about 200 times more massive.

Neutrino A massless, spin-½ lepton often emitted in particle decays.

Neutron A massive particle with no electrical charge. It has roughly the same mass as the proton and is normally found in the nucleus.

New inflationary scenario A version of inflationary theories in which only one domain appears in the GUT freezing.

Nucleosynthesis The process by which nuclei are made from protons and neutrons.

Parity The property of physical systems that tells us whether they are identical when viewed in a mirror.

Parker bound A limit of the number of magnetic monopoles in the galaxy, based on the existence of the galactic magnetic field.

Photon If we wish to think of electromagnetic radiation (from radio wave to light to X rays) as a particle, we call the particle a photon.

Pi meson (Pion) A family of three mesons (mass .14 GeV) that participate in the strong interactions.

Planck time A period of 10^{-43} second after the Big Bang. On time scales shorter than this, the effects of gravity must be included in all physical processes.

Plasma A collection of positive and negative charges that are free to move independently of each other; usually formed by stripping electrons from their nuclei.

Positron The antiparticle of the electron. It has the same mass as the electron but a positive electrical charge.

Proton A massive particle with a positive electrical charge. One of the building blocks of the nucleus.

Quantum mechanical tunneling See tunneling.

Quantum number The set of numbers that specifies the identity of a particle, e.g., spin, electrical charge, strangeness.

Red shift The shift of light emitted by stars in distant galaxies that indicates that they are receding from us.

Relic particles Particles (such as magnetic monopoles and quarks) that did not disappear early in the Big Bang.

Renormalization The process by which apparent infinities arising from calculations of particle interactions are eliminated.

Spontaneous symmetry breaking The process by which a symmetry that holds at high temperature disappears from the system as the temperature is lowered.

Strangeness An internal symmetry analogous to electrical charge. Reactions in which strangeness changes are seen in nature but proceed more slowly than might be expected.

String A massive one-dimensional object that may have been formed 10^{-35} seconds after the Big Bang and subsequently served as the nucleus around which galaxies condensed.

Strong interaction (force) The force responsible for holding particles together in the nucleus or quarks together in the particles.

Superconductors Materials, that at very low temperatures, conduct electricity without any loss.

Supergravity A class of gauge theories of gravitation.

Superparticle The single type of particle that existed before Planck time.

Supersymmetry A symmetry in which bosons and fermions are interchangeable.

Tau meson The third massive lepton, identical to the electron and muon except for its large mass.

Time reversal The mental operation in which we imagine watching a given interaction running backward in time.

Tunneling The process whereby quantum mechanical particles can penetrate barriers that would be insurmountable to classical objects.

Uncertainty principle The quantum mechanical principle that states that the uncertainty in the energy of a system times the uncertainty in the time that the system has had that energy must be greater than a specific number.

Unified field theory Any theory in which two seemingly different interactions are seen to be identical at some deeper level.

Vector bosons Massive spin-1 particles whose exchange generates the weak interaction.

Virtual particle A particle created and reabsorbed in so short a time that violations of energy conservation associated with its mass could not be detected.

Weak interaction (force) The force that governs ordinary radioactive decay.

Weinberg-Salam Model; Weinberg-Glashow-Salam Model The original theory that unified the weak and electromagnetic forces.

X Boson The very massive spin-1 particle associated with GUT and responsible for proton decay.

Index